Logo
AND MODELS OF COMPUTATION

An Introduction to Computer Science

Michael E. Burke
San Jose State University

L. Roland Genise
San Jose Unified School District

Addison Wesley Publishing Company

Menlo Park, California • Reading, Massachusetts
Don Mills, Ontario • Wokingham, England
Amsterdam • Sydney • Singapore • Tokyo
Madrid • Bogotá • Santiago • San Juan

We dedicate this book to

Deanna, Katie, and Kevin

Millie, Livia, Valerie, Roland, Ronnie, and Lisa

M.E.B.
L.R.G.

This book is published by the Addison-Wesley INNOVATIVE DIVISION.

Design and Production by TechArt,
San Francisco, California.

Copyright © 1987 by Addison-Wesley Publishing Company, Inc. All rights reserved. No part of this publication may be reproduced, stored in a retrieval system, or transmitted, in any form or by any means, electronic, mechanical, photocopying, recording, or otherwise, without the prior written permission of the publisher. Printed in the United States of America. Published simultaneously in Canada.

ISBN 0-201-20791-5

ABCDEFGHIJKL-AL-8909876

Contents

Foreword vi
Acknowledgments xiv

Chapter 1 The Logo Calculator

This chapter describes how Logo evaluates expressions and introduces words and lists as the objects used in Logo computations. The primitive operations used with these objects are introduced along with the simplification model for computation. The concept of binding powers is used to clarify the order of evaluation in expressions containing both infix and prefix operations.

1.1	Introduction	1
1.2	Logo Expressions and Their Evaluation	3
1.3	Some Terminology	10
1.4	Computing With Numbers	12
1.5	More on Evaluation of Expressions	15
1.6	Computing With Words	27
1.7	Computing With Lists	30
1.8	Defining Functions and Procedures	34
1.9	Summary of the Simplification Process	39

Chapter 2 The Logo Environment

This chapter introduces the rest of the Logo programming environment as a collection of modules. It is an overview of the system. You may want to spend considerable time investigating particular modules at this time, or you may prefer simply to use it to gain an overall view of Logo.

2.1	Introduction	41
2.2	The Toplevel Module	42
2.3	The Text Screen Module	44
2.4	The Reader Module	49
2.5	The Workspace Module	54
2.6	The Editor Module	59
2.7	The File System Module	63
2.8	The Simplifier Module	67
2.9	Memory Layout	68

Chapter 3 Extending Logo

This chapter describes the graphics capabilities of Logo and introduces recursion. Procedural and functional programming are discussed. **OUTPUT** is contrasted with **SHOW**. The simplification and substitution model is expanded to include user-defined operations and the **REPEAT** procedure.

3.1	Introduction	71
3.2	The Turtle	73
3.3	The Graphics Screen	82
3.4	Defining Procedures for the Turtle	86
3.5	The Simplification of User-Defined Functions and Procedures	89
3.6	The Simplification of Repeat	91
3.7	Recursive Procedures	93
3.8	Procedures versus Functions	97
3.9	Program-Defining Programs	102

Chapter 4 The Ultimate in Computational Power

This chapter introduces the conditional form of control and turns our attention to writing recursive programs. The simplification and substitution model is used extensively to aid in the understanding of recursion. The last three sections describe a significant project involving list processing and graphics.

4.1	Introduction	104
4.2	Recognizers	104
4.3	The **IF** Procedure	109
4.4	Counting Recursion	112
4.5	List Recursion	116
4.6	Piecewise-Defined Functions	121
4.7	Stopping Recursive Turtle Procedures	125
4.8	Mixing Turtles and Lists	128
4.9	Graphing Functions	132
4.10	More on Graphing Functions	138

Chapter 5 Level Diagrams: A Model for Understanding Recursion

The level-diagram model of computation is introduced and used to describe a streamlined execution of Logo programs. It also adds a new medium in which to view recursion. Several new operations are defined to illustrate recursion. Many exercises are provided to give practice in writing recursive functions.

5.1	Introduction	142
5.2	Level Diagrams	142
5.3	Modeling Recursion with Level Diagrams	148
5.4	Some List Utility Functions	153
5.5	Full Recursion	158
5.6	Some Efficiency Considerations	166
5.7	Tail Recursion	173

Chapter 6 Variables

Global and local variables are introduced with a full explanation of the Logo assignment statement. Guidelines for using local and global variables are given along with appropriate warnings. A programming application that maintains a telephone directory is described illustrating the appropriate use of global variables. Writing programs dealing with the mathematical concept of set are presented illustrating the use of local variables. Our models of computation can not deal with the assignment statement and the state diagram model is introduced to cope with the loss.

6.1	Introduction	178
6.2	Global Variables	178
6.3	The Dynamic Scoping Rule	183
6.4	The **MAKE** Statement	189
6.5	Telephone Directory Program	200
6.6	An Interactive Telephone Directory Program	206
6.7	Local Variables	208
6.8	Procedural vs. Functional Programming	210
6.9	Sets as Lists	213

Chapter 7 Programming with Graphical Objects

In this chapter we create programming environments that deal with graphical objects such as multiple turtles and polygons. Property lists are presented as an alternative to global variables. The Towers of Hanoi problem is described in functional terms (returning a list of moves giving a solution) and in terms of programming with graphical objects.

7.1	Introduction	222
7.2	The Turtle as a Collection of Properties	222
7.3	Multiple Turtles	224
7.4	Property Lists	232
7.5	Programming with Multiple Turtles	236
7.6	Geometric Figures as Graphical Objects	238
7.7	The Turtle as a Procedure	242
7.8	Towers of Hanoi	244
7.9	Graphical Solution to the Towers of Hanoi Problem	250

Chapter 8 Building Your Own Computational Environment

This chapter describes the **CATCH** and **THROW** control mechanism of Logo, how it is used in error handling, and how it can be used in the creation of other computational environments. Algorithms for a printer, a scanner, a parser, and a simplifier are given for a rational arithmetic calculator.

8.1	Introduction	259
8.2	The Graceful Return of **EXPLORE.GRAPH**	260
8.3	Going Directly to the Top	264
8.4	Catching Control on the Way to the Top	268
8.5	Logo's Toplevel	275
8.6	The Logo Calculator	277
8.7	Rational-Number Calculator	280
8.8	**RATRUN**: The Rational-Number Expression Simplifier	281
8.9	The Representation of Rational-Number Expressions	285
8.10	**RATREAD**: The Rational-Number Calculator Reader	289
8.11	The Scanner	293
8.12	The Parser	297
8.13	Parenthesized Rational Expressions	301

Chapter 9 Applications and Projects

This chapter presents several projects and numerous exercises dealing with numerical programming, symbolic programming applications, and programming with graphical objects.

9.1	Introduction	305
9.2	A Descriptive Statistics Program	306
9.3	A Calculator for any Number System	309
9.4	Solving Equations	315
9.5	A Set Calculator	319
9.6	Missionaries and Cannibals	322
9.7	Newton's Method	328
9.8	Symbolic Differentiation	330

Appendixes

A.	Some Theory of Computation	335
B.	Glossary	346
C.	Using Apple Logo II	351
D.	Using IBM Logo	355
E.	Logo Primitives For Apple Logo, Apple Logo II, IBM Logo	360
F.	Answers to Selected Exercises	385

Index

Logo Primitives Defined in Chapters 1 through 9	414
General Index	418

Foreword

About the Text

The goals of *Logo and Models of Computation* are to provide an introduction to computer science, to illustrate the relationship of computation to mathematics, to aid in the development of a useful problem solving discipline, and to provide an intellectual framework in which new concepts in computer science can easily be assimilated. The Logo programming language is considered by us to be the ideal vehicle to reach these goals.

The major themes of the text include understanding the difference between functional programming and procedural programming, appreciating and making use of models of computation, thinking about problem solving at various levels of abstraction, and using symbolic computation.

The functional view of programming is characterized by defining operations that produce a result without modifying anything in the computational environment. This style of programming parallels the mathematical concept of function. Furthermore, the mathematical model of simplification and substitution provides a simple way to describe how a computer carries out a computation. This in turn provides an appreciation of the tools of mathematics. The tools of functional programming consist of recursion, conditional branching, and function application. Coupled with the flexibility of lists in the representation of data, the student discovers a conceptually simple form of computation that is very powerful. The concept of recursion is closely related to mathematical induction and Logo provides a very nice environment to discuss mathematical induction.

The procedural view of programming is characterized by defining procedures that execute sequences of statements designed to change the state of one or more objects in the programming environment. The assignment statement is the primary modifier of the world. The substitution and simplification model of computation does not hold up in this setting, and we must turn to another model that views computation as a sequence of states.

Problem solving involves thinking about a problem at several levels of abstraction. Programming can be especially useful in learning this art, but there are two important other ingredients that must be present: The language must support programming at various levels, and the person guiding the exploration of a problem must be able to point out that this is indeed what is happening.

As an example, suppose that the problem concerns creating a game. One aspect is that the roll of a pair of dice will determine a player's options. The ladder of problems associated with the main task is as follows:

1. How is the roll of two dice going to be used?
2. How do you obtain and represent the roll of two dice?
3. How do you obtain and represent the roll of one die?
4. How do you generate a random number in a particular range (an integer between one and six in this case)?
5. How do you generate a random integer in Logo?

The problems are listed in a top-down fashion. Each can be attacked independently of the others. As long as the ladder of subproblems is understood, the order in which they are solved does not matter. In fact each task can be handled by a different person. Person 1 will say to person 2, "Give me two numbers representing the roll of two dice and I'll handle the rest." Person 2 tells person 3, "Give me a way to get the roll of one die, and I'll handle the rest." Person 3 tells person 4, "Give me a way to generate a random number in a particular interval, and I'll handle the rest." Finally, person 4 tells person 5, "Give me a way to generate a random integer in Logo, and I'll handle the rest."

The problem just described will not be solved until all subproblems have been solved, but it does not have to be solved in a top-down fashion or a bottom-up fashion. A middle-out, or "do the even numbered subproblems first," works just as well. In fact, the problem solver is most likely to start with the subproblem he or she understands best.

Logo supports this concept of problem solving better than either FORTRAN or Pascal. BASIC does not support it at all. We have more to say about this later in this preamble.

In the body of the text, we sometimes attack problems at the very bottom. This means we are learning about the level of support provided by Logo. At other times, we attack a problem in the middle. This means that it is only after we solve the problem at hand that we discover the real problem. This process repeated itself several times during the writing of the book. There are a couple of cases where we knew what we were doing from the start. In this case the problem was carefully outlined and implemented in a top-down fashion.

The emphasis of the applications and programs is on writing programs that are symbolic in nature—for example, the manipulation of mathematical objects such as sets, rational numbers, and polygons. In the case of polygons, we do not mean drawing polygons but manipulating them as one would manipulate the turtle. For example, we want to be able to move them around and change their sizes. The properties of polygons are separated from their pictures. They are called *graphical objects*.

We describe programs that deal with nonmathematical objects as well. We experiment with creating a multiple turtle world on top of the single Logo turtle. We solve the Towers of Hanoi problem by writing programs that manipulate disks and towers. We look at the missionaries and cannibals problem in symbolic terms as well.

The applications and programs have been chosen to promote abstract thinking, to broaden the view of computation, and to prepare for further study in LISP, mathematics, artificial intelligence, and computer science.

Who is this Book For?

The material in this book has been used successfully as an introduction to computer science in a course for all students (as well as computer science majors) at San Jose State University and as a text for in-service teacher-training institutes, high school computer courses, and some middle school computing courses.

For the university level course in computer science, the entire book can be covered in an eighteen-week semester. New thinking will need to be done if students have had previous exposure to BASIC and Pascal. Some do not see the light until the last week of the course. For those who choose to go on to a course in LISP and Artificial Intelligence, the rewards are great. The transition is smooth and natural. The usual initial reactions is, "This is just like Logo."

LISP and Artificial Intelligence are currently viewed as advanced courses in most college curriculums. By the time students take these courses, they have been so indoctrinated in the ways of BASIC,

FORTRAN and Pascal that they become lost in a forest of parentheses, and the transition to this rich and exciting field is a difficult one at best. The introductory course and this book were begun because of the near-overwhelming difficulties encountered in the LISP and AI courses. The transition into these courses is now much easier for both students and instructors.

The book has been used in three in-service teacher institutes. The institutes consisted of four weeks of coursework on Logo, mathematics, and problem solving and nine follow-up sessions during the school year. When used in this manner, it is necessary to move quickly (emphasizing the relationship between Logo and mathematics) through the first part of the book and get into the interesting problems in the latter part of the book.

The first thing to admit in this context is that you cannot produce advanced programmers in four weeks. The next thing to admit is that you do not need to. Teachers are in the enviable position of being able to learn while they teach. What they need are reasons to explore and experiment and a solid foundation to lean on. In four weeks they learn to program in Logo and experience a large amount of the material given in the introductory course. If they are motivated they will embark on teaching a variety of projects depending on their individual teaching situation and interests.

We believe the prospective teacher should also be exposed to this material. But this presents a different set of problems and needs another approach. The mathematical background of this group usually varies widely. The emphasis again should be on the relationship of computation and mathematics and problem solving, and the material should be presented in a deliberate fashion.

We do not believe it is important for every teacher (even every mathematics teacher) to be skilled in programming and problem solving, but we do think that every school should have such a teacher. We also feel that every teacher should be exposed to using the computer as a problem-solving tool.

Until we have teachers skilled in using the computer for problem solving, computational literacy will continue to mean: how to turn on a computer, how to insert a diskette in the disk drive, how to be drilled by a computer, how to run a program, and how to write a program. The situation parallels that of mathematics in the elementary school. Since few (if any) elementary schools that have a mathematics teacher, it is not at all surprising that mathematics is viewed by most of the world as the mastery of arithmetic.

The material can be used at the middle school, and the emphasis is the same as at the university level. However, the material is covered at a much slower pace, and time is taken to explain and explore mathematical concepts that are new to these students. It is as much a course in mathematics as in computation.

The material is very appropriate for an introduction to computer science at the high school level. The only difference between high school and college use would be in the rate of coverage.

There are no mathematics prerequisites to the text, but there are two exercises that require trigonometry (they are footnoted), two projects (sections 9.7 and 9.8) that require some calculus, and the last two sections of Chapter 4 assume knowledge of graphing quadratic functions (algebra I). This material can be easily omitted. If a student is taking calculus simultaneously, sections 9.7 and 9.8 should definitely be covered. The presentation of the material is designed to solidify and expand the student's mathematical background, and is valuable in supplementing the mathematics curriculum.

About the Language

The most commonly used programming languages on microcomputers and hence in the public schools, BASIC, and Pascal, do not have the sophistication or the flexibility to develop the kinds of skills we have been talking about. More importantly, these languages encourage primitive levels of problem-solving techniques that tend to limit a person's ability to think on higher levels of abstraction.

The first programming language learned has a strong influence in shaping a person's thinking processes, and higher levels of thinking about a problem are difficult to assimilate if the current frame of reference is a primitive one. Thinking about problems at various levels of abstraction is a major theme of the book. BASIC provides the most primitive level of programming of any of the so-called high-level languages and teaches primitive levels of thinking. Programming in Pascal requires more thinking about writing syntactically correct programs than thinking about how to solve the problem at hand. The data structures of Pascal are also not flexible enough to allow easy representation of symbolic data.

Logo is a direct descendant of a family of languages beginning with LISP in 1959. These languages have been used in the implementation of our most sophisticated programming applications. As a result, Logo has inherited a good deal of the flexibility and sophistication of these languages.

Since many versions of Logo exist (the fundamental concepts are in all of them) and there is no agreed-upon common subset, we are forced to choose a particular version to describe in detail. We have chosen Apple Logo because it can be used with the 64K Apple II Plus computer. It is also appropriate to use the text with Apple Logo II (for Apple IIe and IIc users), IBM Logo (for IBM PC-compatible users), and Atari Logo. In fact, students having these systems at home will encounter no difficulty in adjusting. Appendixes C and D describe the differences encountered if Apple Logo II or IBM Logo is used with the text. Appendix E describes all primitives in Apple Logo, Apple Logo II, and IBM Logo. Experience has shown that students using other versions of Logo (MIT Logo from Terrapin and Krell, or Logo for the Macintosh from Micro-Soft) will face some adjustment of their programs, but without any major difficulty.

We are primarily interested in concepts of computation, and not in memory limitations or execution speed. Someday, Logo compilers will exist and the speed limitation will go away, and next year's hardware will solve today's memory limitations. We also are not overly concerned about the efficiency of our programs, although we do include a few examples illustrating more efficient alternatives. Usually, efficient programs hide the problem we are solving and, as a general rule, efficiency should not be considered until a working program exists. Frequently, inefficiencies and their corrections will jump out at you once a program has been written.

Some Notes on Presentation

1. The **MAKE** statement is not introduced until Chapter 6. We hesitated even to introduce it this early, but some programs would be unnecessarily complicated if we had not. Besides the fact that **MAKE** is widely misunderstood, getting students to fully understand and appreciate recursion is far easier if they do not have **MAKE** at their disposal. This is especially true of students who have been exposed to BASIC. If you allow BASIC programmers the use of **MAKE**, their programs will look just like BASIC programs every time. Furthermore,

simple and elegant models of computation are not possible once the assignment operation is introduced.

2. We use **SHOW** to display values because one never knows what kind of an object (a word or a list) is returned by an expression that is **PRINT**ed. It seems only to add to the confusion about what a particular expression has computed.

3. Confusion always reigns when **OUTPUT** is introduced. Many students have difficulty learning the difference between procedures that **SHOW** a result and functions that **OUTPUT** a result. Students need to be encouraged to define functions that **OUTPUT** a value rather than procedures that **SHOW** a result. If a value is displayed rather than returned, the program cannot be called by another program that would like to use the value in a further computation. The contrast is an effective way to make the concept of function a real one for the student.

4. The student needs to become comfortable with the Logo environment early in the course. Chapter 2 describes the entire Logo environment as a collection of modules, and is presented in one gigantic dump to separate it from the central ideas presented in the rest of the text. In the university-level course it is covered quickly to provide an overview of the environment and is used as a reference for details when needed in a program. It can also be covered in a leisurely manner in a slower-paced course, allowing the student to experiment with and obtain skills in using the editor, writing programs that obtain input from the keyboard, displaying text, and managing the workspace and files. Chapter 2 is also the chapter where the differences in the various versions of Logo are the greatest.

5. In describing the inputs to a procedure or function we use the following notation:

> *number* indicates that the input value must be a number.
> *integer* indicates that the input value must be an integer.
> *word* indicates that the input value must be a word. This includes numbers.
> *list* indicates that the input value must be a list.
> *object* indicates that the input value must be a list or a word.
> *name* indicates that the input value is a word that names something other than a number—for example, a procedure or a function.
> *filename* indicates that the input value is a word that names a file.

6. The **IF** statement is not introduced until Chapter 4. The effect is to delay "interesting" programs until that time. This is a conscious decision because there is already a lot of information that needs to be covered and the concept of conditional control deserves full attention without distracting side issues. Also, the student will more fully appreciate its significance. The instructor may want to introduce it informally at an earlier time if pressed to do so by the students. The assignment statement (**MAKE**) should not be introduced early! In fact, one might want to cover parts of Chapters 7, 8, and 9 before doing Chapter 6.

7. Chapters 7 and 8 are independent of each other. If there is not enough time in the course to cover both chapters, choose which to eliminate based upon the emphasis of the course. Chapter 7 is more appropriate in a class that emphasizes the development of problem-solving skills; Chapter 8 is more appropriate for computer science majors.

8. Chapter 9 consists of projects of varying size and can be covered much earlier. Here is a guide to background needed prior to doing the projects:

   a. Sections 9.2, 9.3, and 9.4 can be covered after Chapter 4.

   b. Section 9.5 requires information from Chapters 6 and 8.

   c. Section 9.6 can be covered after Chapter 6.

   d. Sections 9.7 and 9.8 can be covered after Chapter 4 provided that the student has had some calculus (differentiation).

9. The material in Appendix A serves as an introduction to the theory of computation. It covers computability and proving the correctness of programs. The mathematical concepts introduced and used in this section are informal proofs, including proofs by induction.

10. The word *operation* is commonly associated with the arithmetic operations +, -, *, and /. We use the term *operation* synonymously with the word *function*.

Acknowledgments

We would like to thank the many people who have helped in the development of this book.

Dr. Barbara Pence and Dr. Lynne Gray provided lots of encouragement as well as a lot of constructive criticism of the text. As strong proponents of critical thinking and problem solving, they saw value in the text for mathematics education and insisted on using the text as part of the core for three one-year inservice programs for teachers of mathematics given at San Jose State University from 1984 through 1987. This provided us with the opportunity to use the material in another setting that proved to be extremely satisfying. They also introduced me to a large number of warm, dedicated, and enthusiastic K-14 teachers who were in the process of expanding their own critical thinking and problem-solving skills in mathematics.

Dr. Craig Smorynski wrote the appendix on theory and abstraction. This material explores the theoretical side of computer science using mathematics appropriate for high school students anticipating further study in mathematics or computer science. We regard this material very highly because it links computer science to mathematics, and we expect this bond to become stronger. Dr. Smorynski also aided greatly in making our use of the English language more effective.

Dr. John Mitchem carefully read the text and taught the introductory course in computer science at San Jose State using a preliminary draft. Many of his suggestions on pedagogy have been incorporated.

We are especially indebted to the teachers who have participated in the inservice programs and who keep coming back for more. They provided most of the motivation to finish the book. The students in our classes, in addition to being another testing ground, were a source of ideas as well. They had to cope with the frustrations of trying to make sense out of incomplete, often buggy, preliminary versions of the text. Their input was most valuable. Their questions led to new insights and their expressions told us what worked and what did not—we were all learning together. Specifically, we would like to mention Ronnie Genise who donated uncountable hours in writing, testing, and debugging Logo programs used in our courses as well as giving us his reactions to preliminary versions of the text. He always came up with the answers to our technical problems. We also thank Donna Price for volunteering to put together the appendix of three versions of Logo primitives.

We thank the faculties and administrators of the Department of Mathematics and Computer Science at San Jose State and Steinbeck Middle School for giving us the opportunity and encouragement to test our ideas in the classroom.

Most of all, we appreciate the support and tolerance of our families throughout this writing project.

M.E.B.
L.R.G.

Chapter 1
The Logo Calculator

1.1 Introduction

In this chapter we view Logo as a **calculator**. We usually think of a calculator as a machine that allows us to operate on, or combine, numbers—for example, add, subtract, multiply, and divide numbers quickly and easily. The Logo calculator will do all this, but it also knows about objects called **words** and **lists**. Actually, a number is just a special kind of word.

Our goal is to learn about words, lists, and the operations that are used with these objects and to learn how to interact with the Logo calculator. This means learning to communicate with the machine and finding out how it does its computations. Like any machine, whether it is a car, a bicycle, or a computer, the better you understand it, the better you will use it. Like learning to ride a bicycle or to drive a car, it will feel a little awkward at first, but you will soon find yourself in complete control.

All languages have a basic **structural unit** that is used to communicate a complete message. In the English language, the sentence is the basic structural unit. An English sentence must be constructed by arranging words called verbs, nouns, adjectives, or adverbs in a meaningful way. If we leave out a key ingredient such as the verb, the sentence may be meaningless. If we include words that are unknown to our listener or reader, then the sentence may be misinterpreted or unintelligible to that person.

The basic structural unit in the Logo language is called a **statement**. It begins with a word that names a **procedure**. The named procedure, also called a command, is a request to perform an action such as displaying information on the screen. The name of the procedure is followed by one or more collections of words, called **expressions**, describing the objects used in performing the action. The number of objects needed by a procedure depends on the procedure invoked.

There are no "illegal" statements in the language. Logo will complain if you ask it to do something that it does not know how to do. It will also complain if you fail to provide it with all the information it needs. However, it will attach a meaning (right or wrong) to everything that is typed in. The complaints Logo issues will provide the clues you need to modify a statement so that it communicates the desired request.

Logo will display a greeting, and below the greeting it will display a question mark (?) and a flashing rectangular region called the **cursor**. The question mark is called a **prompt**, a signal to you that Logo is waiting for you to type in a statement. Whatever you type will be displayed at the cursor position, and the cursor will move over one position to the right for each character typed. When the complete request has been typed in, you press the RETURN key to signal that you want Logo to carry it out. Logo does so and then displays the prompt again.

That is, we can describe Logo as a simple repetition of three actions:

Toplevel Logo

1. Display a ? on the screen.

2. Wait for a statement to be typed in from the keyboard.

3. When the RETURN key has been pressed, carry out the requested action.

Here is a sample interaction with Logo:

```
?SHOW 5 + 3
8
?SHOW FIRST [3 9 6]
3
?
```

The first statement requests Logo to display the value of the expression **5 + 3**. The second is a request to display the value of the expression **FIRST [3 9 6]**. The procedure named in each example is **SHOW**. (The procedure name is always the first word in the statement.) A "show statement" displays the value of the expression following the word **SHOW** on the screen.

1.2 Logo Expressions and Their Evaluation

This section defines expressions, discusses the kinds of values returned by expressions, and explains how Logo evaluates expressions. Expressions are common to all programming languages. They are language forms that simplify to a single value. The value is a data object that can be used as an input to another computation.

> A Logo **expression** is a sequence of **words** and **lists** that simplify to a value. The value of an expression is called a **constant**.
>
> A **word** is a sequence of characters.
>
> A **list** is a collection of words and lists that are enclosed in square brackets ([and]).
>
> A **constant** is either a word or a list.

Some examples of Logo expressions and their values are as follows.

Expression: **5 + 3**
Value: **8**

Expression: **FIRST [3 9 6]**
Value: **3**

Expression: **12 / 3**
Value: **4**

Expression **FPUT "FIRST [A B]**
Value: **[FIRST A B]**

Expression: **7 + 3 * 4**
Value: **19**

Expression: **7 = 2**
Value: **FALSE**

Expression: **(7 + 3) * 4**
Value: **40**

Expression: **7 > 2**
Value: **TRUE**

Expression: **7 - 2**
Value: **5**

Expression: **7 < 2**
Value: **FALSE**

Values of expressions are not automatically displayed. The values returned are meant to be used as inputs to a procedure or another operation. For example, you will get the following response if you type in **2 + 3:**

I DON'T KNOW WHAT TO DO WITH 5

This discussion raises more questions than answers. We need to know more about how to write expressions and how Logo evaluates them.

The words in an expression are usually separated by one or more spaces, but certain other characters can also be used to do this. They are: **+, -, *, /, =, <, >, [,], (,** and **)**. Thus, some types of expressions do not need to contain any spaces. An example is the expression **2+3**. The + character separates the two numerals as well as representing the addition operation. Characters that serve as word separators are called **delimiters**.*

Words may not contain any of the delimiting characters (see Exercise 1.2.3 for a special convention that allows for the inclusion of delimiting characters in a word). Some examples of words and nonwords are:

Words	Nonwords
FIRST	**A+B**
X	**[A/2**
:X	**[23]**
"X	**A(AB)**
IF	**(AJB)**
ZZZZ	**"TTTTT$$$]**
236	**2∗3**

A **number** is denoted by the usual base 10 name of the number. Numbers are special kinds of words. That is, a word consisting of a sequence of **Hindu-Arabic numerals** (0, 1, 2, 3, 4, 5, 6, 7, 8, 9) describes a number.

There are two other special words in Logo. The words, **TRUE** and **FALSE** are called **truth values**. The operations **=, <,** and **>** return **TRUE** if the number on the left is respectively equal to, less than, or greater than the number on the right. Otherwise, they return **FALSE**.

---

* The slash character / is not a delimiter in Apple Logo II.

Words are used as the names of operations and constants. For example, the word **SQRT** represents the square root operation, the word **+** represents the addition operation, while the word **26** represents the number twenty-six; the word **TRUE** represents the truth value true.

Logo uses words other than numbers and truth values as constants. That is, it can view a word simply as a sequence of characters and manipulate it by using operations designed to operate on words. For example, **FIRST** is an operation that returns the first character of a word. Since a word like **SQRT** can be viewed in two different ways (the square root operation or the sequence of four characters) we need a way to distinguish these two meanings.

If a word is to be used as a constant other than a number, we precede the word by the double quote character, ". Thus, when we want Logo to return the first character of the word **SQRT** we write

FIRST "SQRT

and when we want to use the square root operation, we do not place the quote character in front of it:

SQRT 25

To further illustrate, consider the following two expressions:

FIRST SQRT 100   (1)
FIRST "SQRT 100   (2)

In (1), **SQRT** is used as the square root operation. It says to return the first character of the word denoting the square root of **100**. Since the square root of **100** is **10**, the value returned by (1) is the word **1**. In (2), **SQRT** is used as a word constant (a sequence of four characters). It says to return the first character of the word **SQRT**. That is, the value is **S**. The number **100** is not part of the expression at all.

A list is a collection of objects. An object in a list may be a word or another list. A list is denoted by enclosing the objects between brackets, [and] and separating the objects inside the brackets by one or more spaces. The following are examples of lists:

1. **[X 10 [A B C]]**
2. **[THIS IS A LIST OF 7 OBJECTS]**

3. [[KATIE 91] [KEVIN 47] [LEFTY 16]]
4. [FD :BLAP + 30]
5. [ ]

Some examples of nonlists are

6.]L I S T
7. SOME LIST]
8. A [BOOK]
9. 2 [+] 3

The objects in a list are called the **elements** of the list. The elements are ordered, so that the first element of the list is the leftmost element, the second element is the one just to the right of the first, and so on.

The lists in (1) and (3) have three elements each. The elements of (1) are the word **X**, the number **10**, and the list **[A B C]**. The elements of the list in (3) are the lists **[KATIE 91]**, **[KEVIN 47]**, and **[LEFTY 16]**. The list in (4) is a four-element list. The list in (5) has no elements and is called the empty list. Notice that no quotation mark is used with the words in the list. The reason for this is that the elements of a list are always treated as constants.

A Logo expression falls into one of two categories: a **constant** or a **combination**. A constant is a word that denotes a number, a word that is preceded by a " character, or a list. For example:

125.735
"ROBOT
""ROBOT
"TRUE
[A [B C] D E]

The expression ""**ROBOT** denotes the six-character word "robot, whereas the expression "**ROBOT** denotes the word robot.

Combinations are defined as follows:

> A **combination** consists of an operation and zero or more expressions describing the constants to be used by the operation.
>
> The constants used by the operation are called the **input values** to the operation.
>
> The expressions describing the input values are called the **input expressions** to the operation.

Examples of combinations are:

1. **3 * 5**
2. **FIRST [3 9 5]**
3. **3 + 5 * 7**
4. **FPUT "FIRST [A B]**

Examples of noncombinations are:

5. **3 ***
6. **FIRST A B]**
7. **3 + 5 * 7 -**
8. **FPUT FIRST [A B]**

In the first example, the operation is ***** and the input expressions are the number constants **3** and **5**. In the second example the operation is **FIRST** and the input expression is the list constant **[3 9 5]**. When **FIRST** is used with a list input, it returns the first element of the list.

In **3 + 5 * 7**, there are two operations. If we ask Logo to evaluate this expression we get **38** as its value. This tells us that the constant **3** and the expression **5 * 7** serve as the input expressions to **+**. In a combination containing more than one operation, there is one that is regarded as the **primary operation** (the last operation to be performed). The other operations are used to describe the input expressions to it. The primary operation in **3 + 5 * 7** is **+**.

The evaluation of a combination depends on which operation is regarded as the primary one. In Section 1.5 we discuss rules that can be used for determining the primary operation. Most of the time we can determine the primary operation by looking at the value produced by Logo.

Example (4) specifies an operation named **FPUT** and the two input expressions, **"FIRST** and **[A B]**. The input expression **"FIRST** indicates that we are using the word **FIRST** as a constant (since it has quotes). **FPUT** (First **PUT**) takes two inputs. The first input may be either a word or a list, whereas the second must be a list. It returns a list identical to the second input with the first input inserted at the front.

The evaluation of a Logo expression is described as follows:

Evaluation of Expressions

The value of a constant is the constant itself.

The value of a combination is the value produced by performing the primary operation on the values of the input expressions.

For example, the values used by **+** in the expression **3 + 5 * 7** are the value of **3** (which is **3**) and the value of **5 * 7** (which is **35**). Thus, the value of the expression **3 + 5 * 7** is **38**.

In **FIRST [3 9 5]**, the input expression to the operation, **FIRST**, is **[3 9 5]**. The value of a list is the list itself (a list is a constant). The value of the expression is the first object in the list, namely, **3**.

In the expression **FPUT "FIRST [A B]**, the value of **"FIRST** is the word **FIRST** and the value of the list **[A B]** is the list itself. Thus, the value is the list, **[FIRST A B]**.

Parentheses can be used to change which operation is the primary one. For example, the value of the expression **3 + 5 * 7** is not changed by writing it as **5 * 7 + 3**. The value in both cases is **38**. We can force the addition to be performed first by writing **5 * (7 + 3)**. That is, parentheses are used to isolate pieces of the expression to show that they should be simplified first. Thus, the expression **5 * (7 + 3)** is evaluated as **50** and the expression **(SQRT 9) + 16** is evaluated as **19**.

Exercise 1.2.1. Using Logo, determine the value of each of the following expressions:

  a. **6 + 5 * 4 / 5**
  b. **6 + 5 * 4 / 5 - 10**
  c. **-10 / 2 / 5 + 1**
  d. **14 = 2 / 7 * 49**
  e. **SQRT 64**
  f. **QUOTIENT 53 7**
  g. **14 = 2 * QUOTIENT 53 7**
  h. **SQRT 9 + 16**
  i. **FIRST [LOGO]**
  j. **FIRST 52 = SQRT 25**
  k. **FPUT "F [A T E]**
  l. **FIRST "CLASS**
  m. **FIRST "CLASS = FIRST [CLASS]**

Exercise 1.2.2. Which of the following are words? For each nonword, tell what delimiters it contains.

  a. **BLUFF**
  b. **BLUFF34**
  c. **BLUFF 46**
  d. **THIS-IS-A%WORD**
  e. **WHATS$THIS?**
  f. **WHERE*IS/FIDO**
  g. **(SECRET)**
  h. **$$$$**
  i. **[[<>]]**
  j. **2=2**

† **Exercise 1.2.3.** Find out if your Logo allows the inclusion of the delimiting characters as characters in a word. If so, what is the convention for doing so?

† **Exercise 1.2.4.** What words other than sequences of numerals are used to represent numbers in Logo?

---

† Whenever this reference mark appears next to exercises, problems, etc., answers are given in the back of this book.

Exercise 1.2.5. Find the values of each of the following expressions.
   a. 3.1416
   b. [VALUE]
   c. FIRST [FIRST]
   d. (FIRST [15 37 24]) + FIRST [6 7 2]
   e. FPUT 2 + 3 [6 7 8 9 10]
   f. FIRST [15 37 24] + FIRST [6 7 2]

Exercise 1.2.6. Determine the primary operation in each of the following expressions by examining the value computed by the Logo calculator:

   a. 15 / 3 + 2
   b. 2 + 3 < 4
   c. FIRST 73 - 4
   d. 3 + FIRST 27
   e. FIRST 27 + 3
   f. FPUT 2 + 3 [X Y Z]
   g. SQRT 9 + 16
   h. 15 / (3 + 2)
   i. 2 + (3 < 4)
   j. (FIRST 73) - 4
   k. FIRST (27 + 3)

1.3 Some Terminology

A combination consists of the primary operation and an input expression describing each input value required by the operation. However, this does not say anything about the order to use in typing them in. That is, all the ingredients to add the numbers 2 and 3 are present in each of the following:

   2 + 3
   + 2 3
   2 3 +

There are three notations for writing combinations.

> **Infix notation:** The symbol representing the operation is placed between the input expressions.
>
> **Prefix notation:** The symbol representing the operation is written to the left of the input expressions.
>
> **Postfix notation:** The symbol representing the operation is written to the right of the input expressions.

The Logo calculator prefers infix notation when arithmetic operations are used and prefix notation for the operations **SQRT, QUOTIENT, FIRST,** and **FPUT**.* In defining a **computational language**, the designer must decide on the notation that will be required in writing expressions.

The designers of Logo decided that infix notation would be preferred for seven operations (+, -, *, /, =, <, and >) and that prefix notation would be required for all other operations. The designers of other programming languages have made different decisions. For example, LISP requires prefix notation exclusively, whereas a language called APL requires postfix notation exclusively. In fact, you can purchase a hand-held calculator that matches any preference you might have. The decision to require infix notation in Logo when arithmetic operations are used was made because that notation is the generally accepted one in mathematics.

Since most operations require either one or two input values, special names have been given to them.

> A **unary** operation is one that requires exactly one input value in order to compute a new value.
>
> A **binary** operation is one that requires exactly two input values in order to compute a new value.

Because it is convenient to refer to the number of input values of an operation, we give this property of operations a name.

---

* Some versions of Logo accept prefix notation for arithmetic expressions as well. For example, both + 2 3 and 2 + 3 are valid.

> The number of input values expected by a Logo operation is called the **arity** of the operation.

Exercise 1.3.1. Identify the type and arity of each of the following operations by placing a check in the appropriate space.

	Operation	Type			Arity			
		Infix	Prefix	Postfix	0	1	2	3 or more
a.	/	✓					✓	
b.	FIRST		✓			✓		
c.	=	✓					✓	
d.	FPUT		✓				✓	
e.	SQRT	✓				✓		
f.	*		✓				✓	
g.	QUOTIENT	✓					✓	
h.	>			✓			✓	

1.4 Computing With Numbers

In this section we describe operations used with numbers and look closer at how Logo evaluates expressions that contain them. We divide them into three categories: (1) the arithmetic operations, (2) the operations that return a truth value, and (3) prefix operations. Operations that return a truth value are called **predicates**.

Arithmetic Operations

number **+** *number*
+ is a binary infix operation that returns the sum of its input values.

number - number
- is a binary infix operation that returns the difference of its input values. The second input is subtracted from the first.

*number * number*
* is a binary infix operation that returns the product of its inputs.

number / number
/ is a binary infix operation that returns the quotient of its inputs. The first input is divided by the second.

Predicates

object = object
= is a binary infix operation that will accept any Logo constants as input values. It returns the word **TRUE** if the first input is equal to the second input. Otherwise, **FALSE** is returned.

number < number
< is a binary infix operation that requires numbers as input values. It returns the word **TRUE** if the first input is less than the second input. Otherwise, **FALSE** is returned.

number > number
> is a binary infix operation that requires numbers as input values. It returns the word **TRUE** if the first input is greater than the second input. Otherwise, **FALSE** is returned.

Prefix Operations

- number
- is a unary prefix operation. The input value must be a number. The negative of its input value is returned.

> **SQRT** *number*
> **SQRT** is a unary prefix operation. It returns a rational approximation to the square root of its input.
>
> **QUOTIENT** *number number*
> **QUOTIENT** is a binary prefix operation. The first input is divided by the second and the integer part of the result is returned. Each input is truncated. That is, the fractional part of an input is discarded if it is not an integer.
>
> **REMAINDER** *number number*
> **REMAINDER** is a binary prefix operation. It returns the remainder when the first input is divided by the second. The fractional part of an input is discarded if it is not an integer.

A source of ambiguity involves the word -. It has been defined as a unary prefix operation as well as a binary infix operation. It is also used to represent negative numbers. That is, we have three different uses for the same symbol. Logo must distinguish among these uses. It does so by examining the context in which it is used. One plausible way to do this is stated next.

> 1. If the - immediately precedes a numeral and follows a delimiter other than right parenthesis, then Logo will view it as a negative number.
>
> 2. If the - immediately precedes a word or left parenthesis, and follows a delimiter other than a right parenthesis, then Logo will view it as the unary operation.
>
> 3. If neither (1) or (2) hold, then - is viewed as binary subtraction.

The following examples show what values are returned if the conventions for determining how - is being used are those just given.

Expression	Value	Convention Used
7 +-2	5	(1)
8 / -SQRT 4	-4.	(2)
7-2	5	(3)
2*(2 + 1) - 2	4	(3)

Examples of the other prefix operations are as follows:

Expression	Value
QUOTIENT 17 3	5
SQRT 49	7.
SQRT 2	1.41421
REMAINDER 17 3	2
REMAINDER 6 3	0

Exercise 1.4.1. Determine the primary operation and the input expressions to the primary operation of each of the following expressions by examining the value computed by the Logo calculator.

† a. QUOTIENT 17 3 + 5
  b. SQRT (9 + 16)
  c. QUOTIENT REMAINDER 39 30 3
† d. (QUOTIENT 17 3) + 5
  e. QUOTIENT 17 + 3 5

1.5 More on Evaluation of Expressions

We have seen several examples of expressions that combine more than one operation together. By examining the value of an expression we are sometimes able to discover what the primary operation is. For example, since the expression

$$2 + 3 * 4 \qquad (1)$$

has value 14 and the expression

$$\text{SQRT } 9 + 16 \qquad (2)$$

has value 5, it is clear that **+** is the primary operation in (1) and that **SQRT** is the primary operation in (2).

We might guess from these examples that the primary operation is the leftmost operation in the expression. But since the value of **3 * 4 + 2** also has value 14, this conjecture must be discarded.

The difficulty arises whenever two operations appear to be "fighting" over the same input value. In (1), **+** and ***** are fighting over the number **3**. In (2) **SQRT** and **+** are fighting over the number **9**. We know that ***** wins the struggle in (1), whereas **+** wins in (2). In fact, ***** will always win a struggle with **+**, and **+** will always win in a struggle with **SQRT**.

These examples illustrate the **binding power** property of operations: ***** is said to have a stronger binding power than **+**, whereas **+** has a stronger binding power than **SQRT**. This can be made clearer by representing the binding power of an operation by a number. The greater the number, the greater the binding power.

Another question we need to answer is: What happens if two operations with the same binding power are fighting over the same number? For example, how is **2 + 3 + 4** evaluated? One answer is that it doesn't really matter because we get the same value regardless of which addition we do first. However, there are some cases where it does matter. For example, consider the following expression:

    2 = 3 = "FALSE

If **2 = 3** is evaluated first, we get **TRUE** as the value of **2 = 3 = "FALSE**, whereas if **3 = "FALSE** is evaluated first, we get **FALSE** as the value. We resolve this difficulty by giving each operation two binding powers: a left and a right binding power. Then, when two operations are fighting over the same input value, the winner is the leftmost operator if its right binding power is greater than the left binding power of the operation on the right.

The numbers representing the left and right binding powers of the operations we have seen are given next.

Operation	Binding Power	
	left	right
*	7	8
/	7	8
+	5	6
-	5	6
=	1	2
<	1	2
>	1	2
All prefix operations	0	4

Logo is not a "standardized" language, and as a result the table of binding powers may not reflect your particular version of Logo. You should attempt to discover the relative binding powers for your version by evaluating a number of expressions—let the results provide the clues.

Since the left binding power of an infix operation is always odd and the right binding power is always even, there is always a clear winner.

Also, notice that prefix operations have lower binding powers than those for the arithmetic operations. Thus, an arithmetic operation always wins a tug of war with a prefix operation. A prefix operation always expects its input values to the right of the operation name, so it is given the lowest left binding power of any operation.

Logo incorporates the binding power properties of operations into a set of agreements called **simplification rules** that can be used in the evaluation of an expression. They are described next. Such rules are needed to define clearly how expressions are simplified. They are also needed if you wish to program a computer to perform the job. However, once you have learned the binding powers for operations, simplification is more easily performed in your head by scanning an expression for any possible conflicts.

Simplification of Constants

If the expression is a constant, it is simplified.

Example 1.5.1
a. 2 => 2
b. [20 30 40] => [20 30 40]

The symbol **=>** is read "simplifies to." For example, **2 + 5 => 7** says "the expression **2 + 5** simplifies to **7**."

Simplification of Prefix Expressions

If the expression begins with a prefix operation, simplify the input expressions and perform the operation.

Example 1.5.2
a. SQRT 6 + 5 * 2
    => SQRT 16
    => 4

b. REMAINDER 8 + 9 SQRT 25
    => REMAINDER 17 SQRT 25
    => REMAINDER 17 5
    => 2

c. QUOTIENT 3 * 5 1 + 1
    => QUOTIENT 15 1 + 1
    => QUOTIENT 15 2
    => 7

d. QUOTIENT 3 * 5 (-3 + 5)
    => QUOTIENT 15 (-3 + 5)
    => QUOTIENT 15 2
    => 7

When simplifying a prefix expression, we must be concerned with determining when one input expression ends and the next begins. This can be determined as follows:

1. The end of the line is reached. (See Example 1.5.2 (a).)

2. An operation (infix or prefix) whose left binding power is less than the right binding power of the prefix operation is seen. (See Example 1.5.2 (b).)

3. A constant is followed by another constant. In this case the second constant denotes the beginning of the next expression. (See Example 1.5.2 (c).)

4. A constant is followed by a left parenthesis. In this case the left parenthesis denotes the beginning of the next expression. (See Example 1.5.2 (d).)

Simplification of Infix Expressions

If the expression begins with a constant and is followed by an infix operation, simplify the right hand input to the operation, perform the operation on its left and right input values and simplify the expression that remains.

Example 1.5.3
a. **6 + 5 * 2**
   => 6 + 10
   => 16

b. **6 + 5 * 2 + 1**
   => 6 + 10 + 1
   => 16 + 1
   => 17

The expression defining the right-hand input of an infix operation begins with the next item and ends by one of the following:

1. The end of the line is reached. (See Example 1.5.3 (a).)

2. An operation (infix or prefix) whose left binding power is less than the right binding power of the infix operation is seen. (See Example 1.5.3 (b).)

3. A constant is followed by another constant. In this case the second constant denotes the beginning of a new expression.

4. A constant is followed by a left parenthesis. In this case the left parenthesis denotes the beginning of a new expression.

Simplification of Expressions within Parenthesis

If the expression begins with a left parenthesis, simplify the expression contained in parentheses and then simplify the expression that remains.

Example 1.5.4
6 * (5 + 2)
=> 6 * 7
=> 42

The following examples further illustrate how expressions are simplified when these rules are employed. In the first two examples, left and right binding powers are added to help clarify what is happening. They are indicated by circled numerals on either side of each operation.

Example 1.5.5 3 * 5 + 4 * 6 / 3

We begin by inserting the left and right binding powers into the expression:

3 ⑦*⑧ 5 ⑤+⑥ 4 ⑦*⑧ 6 ⑦/⑧ 3

Since 8 is greater than 5, 3 * 5 is simplified first:

=> 15 ⑤+⑥ 4 ⑦*⑧ 6 ⑦/⑧ 3

Since 6 is less than 7, **4 * 6** is simplified next:

=> 15 ₍₅₎+₍₆₎ 24 ₍₇₎/₍₈₎ 3

Again, since 6 is less than 7, **24 / 3** is simplified:

=> 15 ₍₅₎+₍₆₎ 8

=> 23

Example 1.5.6   QUOTIENT 3 * 5 + 4 SQRT 36 / 9

₍₀₎QUOTIENT ₍₄₎ 3 ₍₇₎*₍₈₎ 5 ₍₅₎+₍₆₎ 4 ₍₀₎SQRT ₍₄₎ 36 ₍₇₎/₍₈₎ 9

The expression begins with a prefix operation with arity 2. The end of the first input expression is found by scanning right until a binding power less than 4 is encountered. This expression **3 * 5 + 4** is simplified:

=> ₍₀₎QUOTIENT ₍₄₎15₍₅₎+₍₆₎4₍₀₎SQRT₍₄₎36₍₇₎/₍₈₎9
=> ₍₀₎QUOTIENT₍₄₎19₍₀₎SQRT₍₄₎36 ₍₇₎/₍₈₎9

The second input expression is found by continuing to scan right (from the word **SQRT**) until a binding power less than 4 is encountered or until the end of the line is read. Thus, the second input expression to **QUOTIENT** is **SQRT 36 / 9**. It is simplified (using the above rules), leaving us with:

=> QUOTIENT 19 2
=> 9

Example 1.5.7   (2 * 4 - 5) * 6 = 17

=> 3 * 6 = 17
=> 18 = 17
=> "FALSE

Example 1.5.8   Simplify 5 * 3 + 12 * 2 / 4 - 1.

5 * 3 + 12 * 2 / 4 - 1
=> 15 + 12 * 2 / 4 - 1
=> 15 + 24 / 4 - 1

```
=> 15 + 6 - 1
=> 21 - 1
=> 20
```

Example 1.5.9 Simplify 2 + 3 * (1 + (3 * 4) / (6 - 3) * 2) + 4 > 29.

```
   2 + 3 * (1 + (3 * 4) / (6 - 3) * 2) + 4 > 29
=> 2 + 3 * (1 + 12 / (6 - 3) * 2) + 4 > 29
=> 2 + 3 * (1 + 12 / 3 * 2) + 4 > 29
=> 2 + 3 * (1 + 4 * 2) + 4 > 29
=> 2 + 3 * (1 + 8) + 4 > 29
=> 2 + 3 * 9 + 4 > 29
=> 2 + 27 + 4 > 29
=> 29 + 4 > 29
=> 33 > 29
=> "TRUE
```

Using the binding powers and simplification rules just given, we see that the simplification of 2 = 3 = "FALSE proceeds as follows:

```
   2 = 3 = "FALSE
=> "FALSE = "FALSE
=> "TRUE
```

whereas the simplification of 2 = (3 = "FALSE) leads to a different value:

```
   2 = (3 = "FALSE)
=> 2 = "FALSE
=> "FALSE
```

When evaluating expressions beginning with a prefix operation that has arity greater than 1, we need to figure out where the first input expression ends and the second input expression begins. In Example 1.5.6, the binding powers associated with the operations were the decisive factors since the first input expression was an infix expression. This will not help if only prefix operations are used.

Example 1.5.10 Simplify **QUOTIENT REMAINDER 17 6 2**.

Since **QUOTIENT** requires two input expressions, we simplify the next two expressions. The first one is **REMAINDER 17 6** and the second one is the constant **2**. Thus, we have

```
QUOTIENT REMAINDER 17 6 2
=> QUOTIENT 5 2
=> 2
```

In expressions involving several prefix operations (**Example 1.5.11**) it can be difficult to keep track of where we are in the simplification. After performing the simplification in **Example 1.5.11** using the rule on page 18, an alternative method that is less taxing on the human mind is presented.

Example 1.5.11 Simplify **QUOTIENT 17 REMAINDER QUOTIENT SQRT 49 1 4.**

The simplification is performed by simplifying the input expressions to **QUOTIENT** and then applying the **QUOTIENT** operation to those values. The input expressions are:

17

and

REMAINDER QUOTIENT SQRT 49 1 4 (1)

The first input expression is simplified, but the second needs a lot of attention. We analyze it by finding the input expressions to **REMAINDER**:

QUOTIENT SQRT 49 1 (2)

and

4

We must analyze yet another expression, namely, **QUOTIENT SQRT 49 1**. The input expressions to **QUOTIENT** are:

SQRT 49 (3)

and

1

We have finally reached something we can simplify:

SQRT 49 => 6.99999   (**SQRT** returns an approximation; **6.99999** will be displayed as 7.)

Now we can go back to (2) and simplify it:

QUOTIENT SQRT 49 1 => QUOTIENT 6.99999 1 => 6

Now we go back to (1) to simplify it:

**REMAINDER QUOTIENT SQRT 49 1 4
=> REMAINDER 6 4
=> 2**

Finally, we go back to the original expression:

**QUOTIENT 17 REMAINDER QUOTIENT SQRT 49 1 4
=> QUOTIENT 17 2
=> 8**

The complete simplification can be summarized as follows:

**QUOTIENT 17 REMAINDER QUOTIENT SQRT 49 1 4
=> QUOTIENT 17 REMAINDER QUOTIENT 6.99999 1 4
=> QUOTIENT 17 REMAINDER 6 4
=> QUOTIENT 17 2
=> 8**

Many times the simplification process can be carried out more easily by humans if it is analyzed from right to left. Using the last example, we proceed as follows: In scanning the expression from right to left, whenever we see a constant we just move to the left, because there is nothing to do. Constants are already simplified. When we see **SQRT** we know that it requires a single input expression. The input value is just to the right (since everything to our right will be a constant), so we evaluate **SQRT 49**. Thus, we have:

**QUOTIENT 17 REMAINDER QUOTIENT SQRT 49 1 4
=> QUOTIENT 17 REMAINDER QUOTIENT 6.99999 1 4**

Moving left again we encounter the rightmost **QUOTIENT** operation. Its two input values are to the immediate right, so we replace **QUOTIENT 6.99999 1** by its value:

=> QUOTIENT 17 REMAINDER 6 4

Moving left again we find the **REMAINDER** operation. Its two input values are to its right, so we simplify it, obtaining:

   => QUOTIENT 17 2

We next see **QUOTIENT** again, so we do it, arriving at:

   => 8

The first simplification process we did was based on the description of how Logo evaluates expressions. It can get confusing because it is hard to remember where we left off when we finally simplify a piece of the expression. Machines are good at remembering details like this. The second method is more to our liking because we can quickly scan the expression to see what to do next. The first method is much more suited to machines than the second. Likewise, the second method is much better suited to humans.

Exercise 1.5.1. Complete the simplification of each of the following.

a. 10 / 2 * 3 - 9 + 4 / 2   => __5__ * 3 - 9 + 4 / 2
                                   => __15__ - 9 + 4 / 2
                                     => __6__ + 4 / 2
                                   => __6__ + __2__
                                   => __8__

† b. 12 - 3 * 2 + 15 / 3   => 12 - __6__ + 15 / 3
                               => __6__ + 15 / 3
                               => __6__ + __5__
                               => __11__

c. 5 < 7 = "FALSE   => __True__ = "FALSE
                              => __"FALSE__

d. 10 + 3 * 7 = 3 * 7 + 10 = "TRUE

                              => 10 + __21__ = 3 * 7 + 10 = "TRUE
                              => __31__ = 3 * 7 + 10 = "TRUE
                              => __31__ = __21__ + 10 = "TRUE
                              => __31__ = __31__ = "TRUE
                              => __"TRUE__ = "TRUE
                              => __"TRUE__

Exercise 1.5.2. Simplify each of the following expressions, showing all steps.

† a.   10 - (2 * 4) / (5 - 4) * 2 > 0
   b.   3 + 2 * (9 - (3 * 2) / (5 - 2) * 3) + 5
† c.   1 + ((12 + (15 - 7) * 3) / (4 * (9 - 5) + 2)) = 3
   d.   (100 / (3 + (5 * 2) + (14 / 2)) * 5) - 1 = 0

Exercise 1.5.3. Find the primary operation and the input expressions to the primary operation in each of the following and perform the simplification step by step.

† a.   REMAINDER SQRT 100 7
   b.   SQRT REMAINDER 100 7
† c.   -REMAINDER QUOTIENT 25 4 5
   d.   QUOTIENT REMAINDER 15 7 -SQRT 4
   e.   SQRT 1 + 3 * 5 < 15 / REMAINDER 23 4

Exercise 1.5.4. Perform the step-by-step simplification of each of the following expressions.

   a.   -QUOTIENT REMAINDER SQRT 121 6 2
† b.   SQRT -QUOTIENT REMAINDER 19 10 4
   c.   REMAINDER SQRT 81 - QUOTIENT 30 11
† d.   REMAINDER SQRT 81 (-QUOTIENT 30 11)
   e.   QUOTIENT 40 REMAINDER SQRT 25 -2
† f.   QUOTIENT 40 REMAINDER SQRT 25 - 2
   g.   SQRT 1 + 3 * 5 < 15 / REMAINDER 23 4

Exercise 1.5.5. Perform the step-by-step simplification of each of the following expressions.

† a.   SQRT 1 + 3 * 5 < 15 / (REMAINDER 23 4) < 14
   b.   QUOTIENT 17 REMAINDER (- 4 * 3) SQRT 49 + 15 -3
   c.   REMAINDER - 37 SQRT QUOTIENT (4 + (3 * 5)) - 3 2

Exercise 1.5.6. Construct a new binding-power table by interchanging the left and right binding powers of each of the infix operations indicated on page 17 and assigning a right binding power of 10 to prefix operations.

Using the rules for simplification and this new binding-power table, simplify the following expressions.

† a. 2 = 3 = "FALSE
  b. SQRT 9 + 16
  c. QUOTIENT 3 * 5 + 4 SQRT 36 / 4

1.6 Computing With Words

Recall that a word is defined as a sequence of characters. The characters in a word may be any characters you can type on the keyboard. Words are used to represent the names of Logo operations, numbers, and constants that are viewed as a string of characters. Words can be used as building blocks to represent other kinds of objects as well.

We have already seen a few operations that manipulate words. In this section we list them all. They are divided into two categories: **Selectors** are operations that select pieces of words, and **Constructors** are ones that construct new words from old ones.

Selectors

FIRST *word*
FIRST is a unary prefix operation. If the input value is a word, **FIRST** returns the first character of the word.

Examples
?SHOW FIRST "COMPUTER
C
?SHOW FIRST 492
4

LAST *word*
LAST is a unary prefix operation. If the input value is a word, **LAST** returns the last character of the word.

Examples
?SHOW LAST "COMPUTER
R
?SHOW LAST 492
2

BUTFIRST *word*
BUTFIRST is a unary prefix operation. If the input value is a word, **BUTFIRST** returns a new word consisting of all characters in the input word except the first character.

Examples
?SHOW BUTFIRST "COMPUTER
OMPUTER
?SHOW BUTFIRST 492
92

BUTLAST *word*
BUTLAST is a unary prefix operation. If the input value is a word, **BUTLAST** returns a new word consisting of all characters in the input word except the last character.

Examples
?SHOW BUTLAST "COMPUTER
COMPUTE
?SHOW BUTLAST 492
49

Numbers are represented by their base 10 names. Since these names are just words, operations such as **FIRST** can be used with a number input. Whenever the name of a number is used as the input to an operation like **+** or **∗**, the input is treated as if it were a number. Whenever it is used as the input to an operation like **FIRST**, it is viewed as a sequence of characters.

> ### Constructor
>
> **WORD** *word ... word*
> **WORD** is a prefix operation that needs at least two inputs, all of which must be words. It returns a single word constructed by merging its input values together to form a single word.
>
> *Examples*
> **?SHOW WORD "LOGO "COMPUTER**
> **LOGOCOMPUTER**
> **?SHOW WORD 27 492**
> **27492**

We can combine the operations discussed above to perform more complex computations. For example, consider the following interaction with the Logo calculator:

   ?SHOW FIRST BUTFIRST "COMPUTER
   O

Here, **OMPUTER** is returned by **BUTFIRST** and then **OMPUTER** becomes the input to **FIRST**. **FIRST** then returns **O**, which is displayed. Here are some more examples of combinations of word operations:

   ?SHOW BUTFIRST BUTFIRST "LOGO
   GO
   ?SHOW FIRST BUTFIRST BUTFIRST "LOGO
   G
   ?SHOW BUTLAST BUTFIRST "LOGO
   OG

WORD is an operation that can have two or more inputs. This is a problem for evaluation, since it is not known how many inputs to look for. The solution that Logo employs is that when more than two inputs are intended, the user encloses the entire expression in parentheses. For example,

   ?SHOW (WORD "ALL "GOOD "COMPUTERS "NEVER "FAIL)
   ALLGOODCOMPUTERSNEVERFAIL

Parentheses are used for grouping things together when writing any expression. They can also be used to make complex expressions more readable:

```
?SHOW WORD FIRST BUTFIRST "JOHN LAST BUTLAST "JOHN
OH
?SHOW WORD (FIRST BUTFIRST "JOHN) (LAST  BUTLAST
"JOHN)
OH
```

Or, parentheses can be used to force certain operations, like **WORD**, to regard the sequence of expressions following the operation as input expressions. We see examples of two more such operations in the next section.

Exercise 1.6.1. Simplify the following expressions:
† a. BUTFIRST BUTFIRST "ZELDA
  b. FIRST BUTFIRST BUTFIRST "ZELDA
  c. WORD "LOOK BUTFIRST "FFIRST
† d. LAST WORD "BUMBLE "BEE
  e. (WORD FIRST "LUMP "OO LAST "CLUCK)
  f. (WORD "FIRST "LUMP "OO "LAST "CLUCK)

1.7 Computing With Lists

Some of the same operations that were used in computations with words can be used with lists. We describe how these same operations behave when their inputs are lists and also discuss some new operations. Again, they are described as **selectors** and **constructors**:

Selectors

FIRST *list*
FIRST is a unary prefix operation. If the input is a list, the first element of the list is returned.

Examples
```
?SHOW FIRST [A B C D]
A
?SHOW FIRST [[HIYA HIYA] GOODBYE]
[HIYA HIYA]
```

LAST *list*
LAST is a unary prefix operation. If the input is a list, the last element of the list is returned.

Examples
?SHOW LAST [A B C D]
D
?SHOW LAST [[HIYA HIYA] GOODBYE]
GOODBYE

BUTFIRST *list*
BUTFIRST is a unary prefix operation. If the input is a list, a new list whose elements are all of the elements of the input list except the first element is returned.

Examples
?SHOW BUTFIRST [A B C D]
[B C D]
?SHOW BUTFIRST [HELLO [HIYA HIYA] GOODBYE]
[[HIYA HIYA] GOODBYE]

BUTLAST *list*
BUTLAST is a unary prefix operation. If the input is a list, a new list whose elements are all the elements of the input list except the last one is returned.

Examples
?SHOW BUTLAST [A B C D]
[A B C]
?SHOW BUTLAST [HELLO [HIYA HIYA] GOODBYE]
[HELLO [HIYA HIYA]]

Constructors

FPUT *object list*
FPUT is a binary prefix operation. The first input *may* be either a word or a list. The second input *must* be a list. It returns a list whose first element is the first input and whose other elements are the elements of the second input. (Think of it as "FirstPUT.")

Examples
?SHOW FPUT "A [B C D]
[A B C D]
?SHOW FPUT [HIYA HIYA] [HELLO GOODBYE]
[[HIYA HIYA] HELLO GOODBYE]

LPUT *object list*
LPUT is a binary prefix operation. The first input *may* be either a word or a list. The second input *must* be a list. It returns a list whose last element is the first input, and whose other elements are the elements of the second input. (Think of it as "**Last**PUT.")

Examples
?SHOW LPUT "A [B C D]
[B C D A]
?SHOW LPUT [HIYA HIYA] [HELLO GOODBYE]
[HELLO GOODBYE [HIYA HIYA]]

LIST *object ... object*
LIST is an operation that needs one or more inputs, which may be words or lists. It returns a list whose elements are the inputs to the operation.

Examples
?SHOW LIST "A "B
[A B]
?SHOW (LIST [HELLO] [HIYA HIYA] [BYE])
[[HELLO] [HIYA HIYA] [BYE]]
?SHOW (LIST "HELLO [HIYA HIYA] "BYE)
[HELLO [HIYA HIYA] BYE]

If **LIST** is given more than two inputs, then the entire expression must be enclosed in parentheses. If one input is given the entire expression will need to be enclosed in parentheses unless it appears at the end of a statement.

SENTENCE *word ... word*
SENTENCE is an operation that needs at least two inputs, which may be words or lists. If all the inputs are lists, **SENTENCE** returns a list formed by merging the elements of its input lists into a single list. An input that is a word is treated as if it were a single element list.

Examples
?SHOW SENTENCE "A "B
[A B]
?SHOW (SENTENCE [HELLO] [HIYA HIYA] [BYE])
[HELLO HIYA HIYA BYE]
?SHOW (SENTENCE "HELLO [HIYA HIYA] "BYE)
[HELLO HIYA HIYA BYE]

Computing With Lists 33

The difference between **list** and **sentence** can be seen in the following examples:

    LIST "A "B   returns [A B]
    LIST [A] "B   returns [[A] B]
    LIST "A [B]   returns [A [B]]
    LIST [A] [B]   returns [[A] [B]]
    SENTENCE "A "B   returns [A B]
    SENTENCE [A] "B   returns [A B]
    SENTENCE "A [B]   returns [A B]
    SENTENCE [A] [B]   returns [A B]

Exercise 1.7.1. Evaluate the following.

† a. FIRST FIRST [[A [B]] C D]
   b. FIRST BUTFIRST [[A B] C D]
   c. FIRST BUTFIRST BUTFIRST [A B C D]
   d. BUTFIRST [A]
   e. SENTENCE LIST "LEFTY "IS "BRIGHT
† f. SENTENCE FIRST [A B C] LIST "A "B
   g. (SENTENCE "THIS [IS A] "LONG [LIST TO DISPLAY])

Exercise 1.7.2. Evaluate each of the following where possible. For each meaningless expression, indicate what the problem is.

† a. FPUT 2 + 3 [7 9]
   b. FIRST [2 3 5] + 4
   c. SQRT 9 + 16
   d. FIRST 4 + 23
† e. 2 * FIRST [2 3 5] + 4
   f. FIRST [2 3 4] + 5
   g. (FIRST [2 3 4]) + 5
   h. SQRT 25 + FIRST FPUT 2 [3 4 5]
   i. FIRST FPUT 12 [24 36] + -SQRT 144
   j. (FIRST FPUT 12 [24 36]) + -SQRT 144

1.8 Defining Functions and Procedures

One way of increasing the power of Logo is to teach it new operations. This is done by telling Logo the name of the operation, the number of inputs it will expect, and what should be returned as the value.

The value to be returned is described by an expression that indicates how the input values are to be used. Returning the value is caused by executing an **OUTPUT** statement.

> **OUTPUT** *object*
> **OUTPUT** is a unary procedure that can only be used in the body of a function. **OUTPUT** causes the function to return *object*.

The number of inputs is specified by giving a name for each one. The first input name acts as a placeholder for the first input value, the second input name acts as a placeholder for the second input value, and so on.

For example, suppose that we want to define an operation named **THIRD** that returns the third element of a list. The operation requires exactly one input. The name of the input can be any Logo word that is not a number or a truth value. We choose the name **LIST** as the name for the input, since it reminds us that the input value is expected to be a list.

Input names must always be preceded by a colon (:) character (usually read "dots") so that Logo does not confuse input names with names of Logo operations with the same name.

To inform Logo that we wish to define a new operation, we type the word **TO** followed by the name of the new operation and the names of the inputs. For example, to define the operation named **THIRD**, we begin with:

```
?TO THIRD :LIST
>
```

Defining Functions and Procedures 35

After pressing the RETURN key a > character is displayed on the next line. The > is a prompt, or signal, to enter the first of a sequence of statements that will be executed when the new operation is called.

To indicate how to use the real input to **THIRD**, we simply place the input name wherever we want the real input to be used. An expression that computes the third element of the input list is:

FIRST BF BF :LIST

The statement that returns the value of this expression is:

OUTPUT FIRST BF BF :LIST

After typing it in and pressing the RETURN key, we get another > prompt:

```
?TO THIRD :LIST
>OUTPUT FIRST BF BF :LIST
>
```

Since this is the only action we want **THIRD** to perform, we enter the word **END** to tell Logo there are no other statements in the definition of **THIRD**:

```
?TO THIRD :LIST
>OUTPUT FIRST BF BF :LIST
>END

THIRD DEFINED
?
```

The collection of statements entered between the line beginning with **TO** and the line containing the word **END** is called the **body** of the new operation. In the case of **THIRD**, there is just one statement in the body.

Every time you press RETURN, a new > prompt appears. It tells you that Logo is waiting for the next statement in the sequence making up the body or the word **END**. The word **END** tells Logo that there are no more Logo statements in the body.

The reply

THIRD DEFINED

36 *The Logo Calculator*

is confirmation that the definition of **THIRD** has been added to Logo's collection of known operations.

A new piece of the Logo language has been introduced here, namely, dots followed by a word. This new object is called a *variable*.

> A word preceded by dots is called a **variable**.

One use of variables is to act as placeholders, indicating how many inputs are needed in a user-defined operation and how they are to be used when the operation is evaluated. When used in this way, variables are called **input variables**.

Once a new operation is defined, we can use it as if it were one of the built-in ones. For example, we can now use **THIRD** to extract the third element of a list:

```
?THIRD [A B C D E]
YOU DON'T SAY WHAT TO DO WITH C!
?SHOW THIRD [A B C D E]
C
?SHOW THIRD (LIST "A [B C] [D E] F)
[D E]
?SHOW THIRD "ROBOT
B
?SHOW FPUT THIRD [A B C D E] [X Y Z]
[C X Y Z]
```

The built-in operations are called **primitive** operations, whereas the ones that the user defines are called **user-defined** operations. The simplification of an expression that contains a user-defined operation is a straightforward extension of the simplification rules given earlier.

When a user-defined operation is called, Logo first substitutes the real inputs for each occurrence of the corresponding input variable in the body. The first real input is substituted for each occurrence of the first input variable. The second real input (if there is one) is substituted for the second input variable, and so on. Logo then executes the statements in the body sequentially. An **OUTPUT** statement is simplified by replacing it with its input expression. The simplification of **THIRD [A B C D E]** can be described as follows:

THIRD [A B C D E]
=> OUTPUT FIRST BF BF [A B C D E]
=> FIRST BF BF [A B C D E]
=> FIRST BF [B C D E]
=> FIRST [C D E]
=> "C

The simplification of **FPUT THIRD [A B C D E] [X Y Z]** is given by:

FPUT THIRD [A B C D E] [X Y Z]
=> FPUT OUTPUT FIRST BF BF [A B C D E] [X Y Z]
=> FPUT FIRST BF BF [A B C D E] [X Y Z]
=> FPUT FIRST BF [B C D E] [X Y Z]
=> FPUT FIRST [C D E] [X Y Z]
=> FPUT "C [X Y Z]
=> [C X Y Z]

The expression, **THIRD [A B C D E]** is replaced by a copy of the body, where the real input has been substituted for the input variable. The **OUTPUT** statement is then replaced by its input expression and the simplification continues according to the simplification rules given earlier.

An operation is also called a **function**. Functions differ from procedures in that a function computes (returns) a value that can be used as an input to some other procedure or function. For example, compare the following definition with the definition of **THIRD**.

?TO SHOWTHIRD :LIST
>SHOW FIRST BF BF :LIST
>END

SHOWTHIRD DEFINED
?

Now, execute

?SHOW THIRD [A B C D E]
C

and

```
?SHOWTHIRD [A B C D E]
C
```

There does not seem to be much difference between **THIRD** and **SHOWTHIRD** until we try,

```
?SHOW FPUT THIRD [A B C D E] [X Y Z]
[C X Y Z]
```

and

```
?SHOW FPUT SHOWTHIRD [A B C D E] [X Y Z]
C
NOT ENOUGH INPUTS TO FPUT
```

The problem with the second example is that **SHOWTHIRD** does not return a value that can be used as an input to **FPUT**. Thus, returning a value and displaying a value on the screen are very different kinds of actions.

Exercise 1.8.1. Define the following functions.

† a. Define a function named **FOURTH** that takes a word as its only input and returns the fourth character of the word.
  b. Define a function named **SEVENTH** that takes a word as its only input and returns the seventh character of the word.
  c. Define a function named **REPLACE.BY.A** that takes a word as its only input and returns a new word identical to the input word except that the first character is replaced by the letter **A**.
  d. Define a function named **REPLACE.BY.C** that takes a word as its only input and returns a new word identical to the input word except that the third character is replaced by the letter **C**.
† e. Define a function named **DELETE.THIRD** that takes a word as its only input and returns a new word identical to the input word except that the third character is deleted.

Exercise 1.8.2. Define the following functions.

† a. Define a function named **FOURTH.ITEM** that takes a list as its only input and returns the fourth element of the list.
  b. Define a function named **SEVENTH.ITEM** that takes a list as its only input and returns the seventh element of the list.

c. Define a function named **REPLACE.ITEM.1** that takes a list as its only input and returns a new list identical to its input except that the first element is replaced by the word **FIRST**.

† d. Define a function named **REPLACE.ITEM.3** that takes a list as its only input and returns a new list identical to its input except that the third element is replaced by the word **THIRD**.

e. Define a function named **DELETE.THIRD** that takes a list as its only input and returns a new list identical to its input except that the third element is deleted.

1.9 Summary of the Simplification Process

The Logo language consists of statements and expressions. Expressions describe the input values to procedures and functions.

Statements and Expressions

A statement consists of the name of a procedure and its input expressions.

An expression is any one of the following:

1. A constant
2. A combination
3. An expression enclosed in parentheses

A constant is either a word or a list.

A combination consists of the name of a function and its input expressions.

Additional functions and procedures may be defined by the user, who must specify the following three things.

> The three components of a user-defined function or procedure are
>
> 1. Its name
> 2. The names of the input variables (if any)
> 3. The body

Statements are executed by carrying out the indicated procedure on the objects specified by the simplification of the procedure's input expressions. The rules for simplifying expressions are as follows.

> **Simplification of Logo Expressions**
>
> If the expression is a constant it is simplified.
>
> If the expression begins with a Logo primitive function, simplify the input expressions and perform the operation.
>
> If the expression begins with a user-defined function, simplify the input expressions and replace the expression with the body of the function after substituting the input values for the corresponding input variables.
>
> If the expression begins with a constant and is followed by an infix operation, simplify the right-hand input to the operation, perform the operation on its left and right input values, and simplify the expression that remains.
>
> If the expression begins with a left parenthesis, simplify the expression contained in parentheses and then simplify the expression that remains.

Chapter 2
The Logo Environment

2.1 Introduction

The entire Logo environment is pictured in Figure 2.1. It consists of nine modules, each of which is designed to handle specific tasks. Each task is performed by a Logo primitive operation or procedure. Each module consists of a collection of one or more related objects and the primitives that use the objects to perform their tasks.

```
                    ┌─────────────┐
                    │  TOPLEVEL   │
                    └─────────────┘

  ┌─────────────────────┐      ┌─────────────────────┐
  │ READER              │      │ SIMPLIFIER          │
  ├─────────────────────┤      ├─────────────────────┤
  │ READLIST            │      │ RUN list            │
  │ READCHAR            │      └─────────────────────┘
  │ KEYP                │      ┌─────────────────────┐
  │ PADDLE number       │      │ TEXT SCREEN         │
  │ BUTTONP number      │      ├─────────────────────┤
  └─────────────────────┘      │ CLEARTEXT           │
                               │ PRINT object        │
                               │ SHOW object         │
  ┌─────────────────────┐      │ TYPE object         │
  │ FILE SYSTEM         │      │ SETCURSOR object    │
  ├─────────────────────┤      │ CURSOR              │
  │ SAVE word           │      └─────────────────────┘
  │ LOAD word           │      ┌─────────────────────┐
  │ ERASEFILE word      │      │ WORKSPACE           │
  │ CATALOG             │      ├─────────────────────┤
  └─────────────────────┘      │ DEFINE word list    │
                               │ ERASE word          │
  ┌─────────────────────┐      │ ERPS                │
  │ EDITOR              │      │ PO word             │
  ├─────────────────────┤      │ POPS                │
  │ EDIT                │      │ POTS                │
  │ EDIT word           │      │ TEXT word           │
  │ EDIT word           │      │ DEFINEDP            │
  └─────────────────────┘      └─────────────────────┘

  ┌─────────────────────┐      ┌─────────────────────┐
  │ TURTLE              │      │ GRAPHICS SCREEN     │
  └─────────────────────┘      └─────────────────────┘
```

Figure 2.1

All the modules in Figure 2.1 are described in this chapter with the exception of the **Turtle** and the **Graphics Screen**. They will be described in Chapter 3.

This modular description provides a logical view of the entire language as a small number of building blocks. If we want to know more about the tasks performed by one of the modules, we need only look at that particular module. For example, if we want to find out how to display text on the screen, we examine the **Text Screen** module for primitives that will help.

The **Toplevel** module is responsible for getting a statement from the user and setting in motion the chain of events that will handle the request. It calls on the **Simplifier** to simplify the input expressions to the requested action specified in the statement, and executes the simplified statement.

A **simplified statement** consists of the name of the desired procedure and the actual input values required to perform the task. For example, the statement **SHOW 5** is the simplified form of the statement **SHOW 2 + 3**.

2.2 The Toplevel Module

The **Toplevel** module performs the actions described at the beginning of Chapter 1:

1. Display the ? prompt.
2. Read a statement.
3. Perform the desired action.

Like a good top-level manager, it performs these tasks by passing them off to other modules. It then repeats the actions all over again. The desired tasks are performed by the Logo primitives **TYPE, READLIST,** and **RUN**. Thus, the **Toplevel** module can be described by the following Logo procedure:

```
TO TOPLEVEL
TYPE "?
RUN READLIST
TOPLEVEL
END
```

The body of the procedure consists of three statements.

TYPE is a procedure that belongs to the **Text Screen** module. The job of the **Text Screen** module is to display words and characters on the screen. **TYPE** displays its input value (the ? character in this case) on the screen. **TYPE** differs from **SHOW** in that the cursor is left immediately to the right of what gets displayed when **TYPE** is called. If **SHOW** were used, the cursor, and hence the next characters displayed, would be placed on the following line of the screen.

READLIST is a function that belongs to the **Reader** module. The **Reader**'s job is to get information from the user via the keyboard, game paddles, or other devices supported by the computer. **READLIST** returns a list of words and lists entered by the user. This list becomes the input value to the **RUN** procedure.

RUN is a procedure that belongs to the **Simplifier** module. The **Simplifier**'s job is to simplify the input expressions of a statement and execute the result. For example if the user enters **SHOW 2+3**, after the prompt has been displayed, the **Reader** will return the list [SHOW 2 + 3]. The **Toplevel** module then calls on **RUN** to obtain and execute the simplified statement, **SHOW 5**.

The input to **RUN** must be a list containing one or more Logo statements.

The last statement in the **Toplevel** procedure is a call to itself. Thus, the same three statements are executed again and again.

Exercise 2.2.1. Execute the following statements:

    a. TYPE "COMPUTER
    b. TYPE "S
    c. TYPE "ARE
    d. SHOW "DUMBO
    e. SHOW READLIST SHOW 2+3
    f. RUN [SHOW 2+3]
    g. RUN [FPUT FIRST [A B C] [X Y Z]]
    h. RUN READLIST
    i. RUN [READLIST]
    j. SHOW RUN [READLIST]

Exercise 2.2.2. Define a procedure named **MYTOP** that acts just like Logo's top-level procedure except that the prompt is changed to **FIREAWAY:**. That is, after defining **MYTOP**, the following interaction should occur:

```
?MYTOP
FIREAWAY: SHOW 2 + 3
5
FIREAWAY: SHOW FIRST [FIREAWAY BUDDY]
FIREAWAY
FIREAWAY:
```

Exercise 2.2.3. Modify **MYTOP** so that it displays the message "go for it" after completing the execution of the user's request.

2.3 The Text Screen Module

The **Text Screen** module consists of a collection of character positions, the position of the cursor, and the primitives that deal with these objects.

The character positions are viewed as 24 lines of 40 or 80 characters per line. They contain the text that is currently being displayed on the computer's CRT (cathode ray tube). We refer to this object as the **text screen**.

The **text screen** is used to display statements entered by the user and the values produced (provided that a **SHOW, PRINT,** or **TYPE** statement was entered). The cursor marks the position on the screen at which the next character will be displayed.

For example, the text screen in Figure 2.2 indicates that the statement **SHOW 2 + 3** has been typed in at the top of the screen, the result has been displayed, and that the reader is waiting for a new statement to be typed in.

The Text Screen Module 45

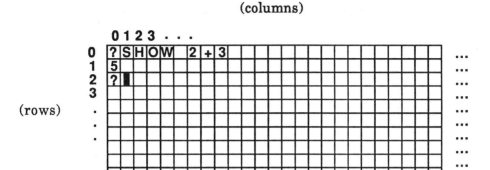

Figure 2.2 Text Screen

The **Text Screen** module is represented by two lists and the primitive functions and procedures that deal with the text screen and the cursor. One list represents the collection of character positions on the text screen itself. It is a list containing 24 elements. Each element represents a row of 40 (or 80) characters. That is, each element is a list of characters whose length is no greater than the length of a row.

The second list represents the cursor. It is a list of two numbers indicating the column and row numbers of the cursor's current position. Figure 2.3 summarizes the objects and primitives associated with the **Text Screen** module.

```
┌─────────────────────────────────────────────────────────────┐
│ ┌──────────────┐                                            │
│ │ Text Screen  │                                            │
│ └──────────────┘                                            │
│                                                             │
│      OBJECTS                                                │
│                                                             │
│        Text Screen:                                         │
│         [ [. . . characters in row 1   . . .]               │
│           [. . . characters in row 2   . . .]               │
│           [. . . characters in row 3   . . .]               │
│                        . . .                                │
│                        . . .                                │
│                        . . .                                │
│           [. . . characters in row 24  . . .] ]             │
│      Cursor Position: [CR] (column and row numbers)         │
│      PRIMITIVES                                             │
│                                                             │
│        Procedures                  Functions                │
│                                                             │
│        CLEARTEXT                   CURSOR                   │
│        PRINT object...object                                │
│        SHOW object                                          │
│        TYPE  object...object                                │
│        SETCURSOR list                                       │
│                                                             │
└─────────────────────────────────────────────────────────────┘
```

Figure 2.3 The Text Screen Module

The text screen and cursor in Figure 2.2 are represented by the following lists:

[[? S H O W 2 + 3] [5] [?] [ ] ... [ ]]
represents the characters currently occupying positions on the text screen.

[1 2]
represents the position of the cursor.

The primitives that deal with the text screen are described below.

Characters can be placed anywhere on the text screen by using the **SETCURSOR** procedure to position the cursor and then calling on a procedure that causes text to be written on the screen (**PRINT, TYPE**, or **SHOW**). The **Text Screen** module is responsible for placing characters in the appropriate locations in the list representing the screen and updating the list representing the cursor.

Text Screen Primitives

Procedures

CLEARTEXT
CLEARTEXT is a procedure that causes the text screen to be cleared of all characters.

PRINT *object ... object*
PRINT is a procedure that takes an arbitrary number of words and lists as inputs. The words and lists are put on the text screen starting at the current cursor position. If the input is a word the word delimiter is not displayed. If the input value is a list, the outside list delimiters ([and]) are not displayed. The cursor is placed at the beginning of the next line.

Example
?(PRINT [GREETINGS FROM] [THE TEXT SCREEN])
GREETINGS FROM THE TEXT SCREEN

SHOW *object*
SHOW is a procedure that takes a word or list as its only input. It is identical to **PRINT** except that outside list delimiters are displayed.

Example
?SHOW [GREETINGS FROM THE TEXT SCREEN]
[GREETINGS FROM THE TEXT SCREEN]

TYPE *object ... object*
TYPE is a procedure that takes an arbitrary number of words and lists as inputs. It behaves like **PRINT** except that there is no space between the objects placed on the screen and the cursor is left at the position immediately following the last character displayed.

Example
?(TYPE [GREETINGS FROM] [THE TEXT SCREEN])
GREETINGS FROMTHE TEXT SCREEN

> **SETCURSOR** *list*
> **SETCURSOR** is a procedure that takes a list of two numbers as its only input. The cursor is moved to the indicated position. The first number in the list is the column number (0 to 39 or 79) and the second is the row number (0 to 23).
>
> *Example*
> **?SETCURSOR   [10 15] SHOW   [TEXT SCREEN GREETINGS]**
> (The list will be displayed at column 10 row 15.)
>
> *Function*
> **CURSOR**
> **CURSOR** returns the current position of the cursor as a list of two numbers indicating the column and row numbers.
>
> *Example*
> **?SHOW CURSOR**
> **[43 15]**

Example 2.3.1 Consider the following sequence of statements:

   ?SETCURSOR [30 10] PRINT "HI.THERE
   ?SETCURSOR LIST (FIRST CURSOR) + 2 (LAST CURSOR) PRINT "HI

The first example causes the word **HI.THERE** to be displayed on the screen starting at column 30 row 10.

The second example causes the word **HI** to be displayed on the screen starting two columns to the right of the current cursor position.

We can define our own procedures to manipulate the text screen as well. For example, to display the first three characters of a word diagonally on the screen starting at a particular location we could define the following procedure:

```
TO TYPE.DIAGONALLY :WORD :LOCATION
SETCURSOR :LOCATION
TYPE FIRST :WORD
SETCURSOR LIST 1 + FIRST :LOCATION 1 + LAST :LOCATION
TYPE FIRST BF :WORD
SETCURSOR LIST 2 + FIRST :LOCATION 2 + LAST :LOCATION
TYPE FIRST BF BF :WORD
END
```

Exercise 2.3.1. Define a procedure named **TYPE.DOWN** that takes a word and a list representing a character position as input and displays the first three characters of the word input vertically down starting at the specified character position.

Exercise 2.3.2. Define a procedure named **MY.NAME** that displays your name and address in the middle of the screen. Each line should be centered.

† **Exercise 2.3.3.** Define a procedure named **ASSIGNMENT.HEADER** that takes a number representing an assignment number and a list of exercise numbers as its inputs. It displays this information along with your name, class, and term at the top of the screen. For example, **ASSIGNMENT.HEADER 3 [2.2.1, 2.2.2, 2.2.3]** results in:

Ronnie Hacker
Math 45, Fall 1986
Assignment: 3, Problems: 2.2.1, 2.2.2, 2.2.3

2.4 The Reader Module

The **Reader** module is responsible for obtaining input from the user through any of a number of devices. The keyboard is the device used most often. Other devices used to provide input to programs might include game paddles or a mouse. Logo includes appropriate functions to gather and return input from each device. We describe how to get information from the keyboard as well as from game paddles.

The primary use of the keyboard is to type in statements for Logo to execute, but it may also be called upon to enter data needed by a program. There is no real difference in these two tasks as far as the **Reader** module is concerned. In the first case, the **Toplevel** module executes **READLIST** to obtain a statement to execute. In the second, a program may execute **READLIST** to obtain needed data. For example, consider the following program:

```
TO MOVECURSOR
TYPE [ENTER THE DESIRED CURSOR POSITION:]
SETCURSOR READLIST
TYPE [IS THIS WHERE YOU WANT THE CURSOR?]
END
```

This procedure displays some text that prompts the user to type in numbers describing a cursor location. The user is expected to type in two numbers separated by a space. When the RETURN key is pressed **READLIST** returns a list containing the two numbers. The cursor is moved to the location described by the input list. Finally, the question is placed on the text screen starting at the location specified by the user.

Whenever a key (including the RETURN key and space bar) is pressed, that character is placed in an object called the **input stream**. This ensures that it won't get lost if a **READLIST** or a **READCHAR** statement has not been executed yet. In addition, the **Reader** executes a **TYPE** statement so that each character is displayed on the text screen.

The input stream can be thought of as a list of up to 255 characters. When keys are pressed, the specified characters are placed at the end of the input stream and **TYPEd** on the text screen. When characters are read, the characters at the beginning of the input stream are removed.

If more than 255 characters are entered without reading them, the computer will issue a beeping noise to inform you of this condition. When the input stream becomes full, all subsequent characters typed will be lost.

When **READLIST** is executed, nothing will be read from the input stream until a RETURN character has been added to it. Thus, we have a chance to change what has been typed in so far. Logo provides several editing commands to provide this utility.

Since most keys will insert a character in the line buffer (memory used to temporarily store data), all the editing commands are provided by control characters. Control characters are entered by pressing a key while holding the CONTROL key down. Control characters are not entered into the input stream.

The **Reader** module is described in Figure 2.4.

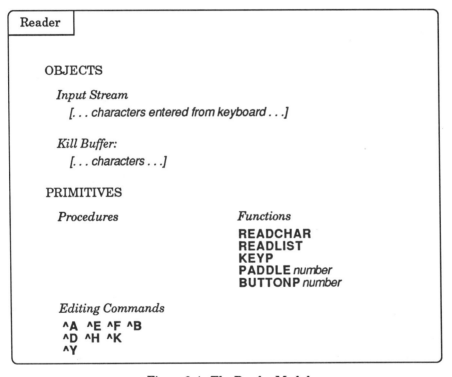

Figure 2.4 The Reader Module

The following primitives are provided by Logo to read characters in the input stream.

Functions for Reading the Input Stream

READLIST
READLIST is a function with no inputs. A list of all the words and lists in the input stream up to the first RETURN character is returned. If no RETURN character is found, **READLIST** waits until a RETURN character is entered.

READCHAR
READCHAR is a function with no inputs. The next character in the input stream is returned. If the input stream is empty, **READCHAR** waits until a key is pressed.

KEYP
KEYP is a function with no inputs. It returns **TRUE** if there is a character in the input stream and **FALSE** otherwise.

A game paddle consists of a knob that can be rotated and a button that is either in or out. The primitives that deal with reading the game paddle read the position of the knob or determine whether the button is in or out.

Functions for Reading the Game Paddles

PADDLE *number*
PADDLE is a function with a number indicating which paddle to read as its only input. It returns a number between 0 and 255 indicating the position of the paddle knob.

BUTTONP *number*
BUTTONP is a function with a number indicating which paddle to read as its only input. It returns **TRUE** if the button is down and **FALSE** otherwise.

The following procedure displays the result of reading the positions of the paddle knobs continuously:

```
TO PADDLEPRINT
PRINT PADDLE 0
PRINT PADDLE 1
PADDLEPRINT
END
```

While a small number of editing commands may be associated with particular keys on the keyboard, we describe all the editing commands available by their control key counterpart.

There are four editing commands that are used to move the cursor to any character in the input stream:

CONTROL-A moves the cursor to the beginning of the stream.

CONTROL-E moves the cursor to the end of the stream.

CONTROL-F moves the cursor forward (right) one character.

CONTROL-B moves the cursor back (left) one character.

There are three editing commands that are used to delete one or more characters from the input stream:

CONTROL-D deletes the character under the cursor.

CONTROL-H deletes the character to the left of the cursor.

CONTROL-K moves all characters under and to the right of the cursor into the kill buffer.

There is an additional command that provides for copying text:

CONTROL-Y copies the contents of the kill buffer into the input stream. Every time a RETURN character is entered, the contents of the input stream are copied into the kill buffer. This allows you to execute the same statement again without having to retype it. You may also edit it (before pressing RETURN) to provide a different instruction.

Exercise 2.4.1. Define a procedure named **PUT.STAR** that requests a column and row number of a character position on the text screen from the user and then displays an asterisk, *, at that location. Since whatever is typed by the user is displayed at the current cursor position, use the first line on the screen to prompt for the requested information.

Exercise 2.4.2. Define a procedure named **STAR.DRAW** that calls **PUT.STAR** over and over again.

2.5 The Workspace Module

When user-defined functions and procedures are created, they are stored in an object called the **workspace dictionary**. The **Workspace** module consists of this dictionary and the primitives that modify the dictionary.

The dictionary is represented by a list of entries. Each entry in the dictionary is a two-element list containing the name of a function or procedure and its definition. The definition consists of a list of the input variable names for the operation and a list for each statement in the body.

The primitives that deal with the workspace dictionary are used to add or modify an entry in the dictionary, to display the contents of the dictionary, and to return a copy of a definition contained in the dictionary. The **Workspace** module is summarized in Figure 2.5 and the primitives that deal with the workspace dictionary are described in the definition box on page 56.

Here are some examples illustrating the use of these primitives:

```
?DEFINE "SECOND [ [OBJECT] [OUTPUT FIRST BF :OBJECT]]
?DEFINE "FIRST.TWO
    [[OBJECT][OUTPUT SE FIRST :OBJECT FIRST BF :OBJECT]]

?PO "SECOND
TO SECOND :OBJECT
OUTPUT FIRST BF :OBJECT
END

?POPS
TO SECOND :OBJECT
```

OUTPUT FIRST BF :OBJECT
END

TO FIRST.TWO :OBJECT
OUTPUT SE FIRST :OBJECT FIRST BF :OBJECT
END

?POTS
TO SECOND :OBJECT
TO FIRST.TWO :OBJECT

?ERASE "FIRST.TWO

?POTS
TO SECOND :OBJECT

?SHOW TEXT "SECOND
[[OBJECT] [OUTPUT FIRST BF :OBJECT]]

```
Workspace

    OBJECTS

        Dictionary
        [SECOND
          [ [ OBJECT ]
            [ OUTPUT FIRST BF :OBJECT ] ] ]

        [SHOW.FIRST.TWO
          [ [ OBJECT ]
            [ OUTPUT SE FIRST :OBJECT FIRST BF :OBJECT ] ] ]

    PRIMITIVES

        Procedures              Functions
        DEFINE word  list       DEFINEDP word
        ERASE object            TEXT word
        ERPS
        PO object
        POPS
        POTS
```

Figure 2.5 The Workspace Module

Workspace Primitives

Procedures

DEFINE *word list*
DEFINE requires two inputs and is used to add a new entry to the workspace dictionary. The first input must be a word that indicates the name of the operation to be added or modified. The second input is a list whose elements are a list of input variable names and lists of Logo statements making up the body of the new operation.

ERASE *object*
ERASE takes a word or a list of words as its only input. It deletes the procedure(s) and function(s) named by the word(s) from the dictionary.

ERPS
ERPS requires no inputs. It erases all procedures and functions from the dictionary.

PO *object*
PO requires a word or a list of words as its only input. It displays the **TO** form of the named functions and procedures.

POPS
POPS requires no inputs. It displays the **TO** forms of all functions and procedures in the dictionary.

POTS
POTS requires no inputs. It displays the names and input variables of all functions and procedures in the dictionary.

Functions

TEXT *word*
TEXT takes a word as its only input. It returns the definition of the named procedure or function as a list

> whose elements are the list of input variable names and a list for each statement making up the body of the definition.
>
> **DEFINEDP** *word*
> **DEFINEDP** takes a word as its only input. It returns **TRUE** if there is a function or procedure named by the input word. Otherwise it returns **FALSE**.

New functions and procedures can be defined using the scheme described in Chapter 1 or by using the **DEFINE** procedure. You will quickly find that using the **TO** form is a lot friendlier, because **DEFINE** requires you to package everything up into lists. Furthermore, it becomes awkward when you define operations that contain more than 255 characters, since the reader will not read more than 255 characters at a time.

The **TO** form is handled differently by the **Toplevel** module. After getting the line starting with **TO** from the **Reader**, the **Toplevel** module enters a different mode. It does not send the **TO** form to the **Simplifier** for execution. In the **TO** mode, the name of the operation is remembered and the names of the input variables are stuffed into a list. The list of variable names is put onto the front of another list, which will also be used as a receptacle for the statements contained in the body.

As you saw in Chapter 1, you are prompted for each statement (line) in the body. The **Toplevel** module executes a **READLIST** and **LPUTs** the result (a statement in the body) onto the list whose first element is the list of variable names.

A list containing the definition of the function or procedure is being built. When a line containing the word **END** is encountered, a **DEFINE** statement is constructed and **RUN**. The inputs to **DEFINE** are the name of the operation and the list containing the input variables and body.

All the editing commands described for the **Reader** module can be used when defining an operation using the **TO** mode, but once you press RETURN, you cannot bring that line back to edit it. Thus, after typing in one line, look at it carefully before going on to the next line.

TO and **END** are not Logo primitives. **TO** can be used only at Logo's **Toplevel** (in response to the **?** prompt) and cannot be used as part of the body of the definition. We can, however, use a **DEFINE** procedure in the body of a new definition. That is, we can write (and do in the next chapter) programs that write (**DEFINE**) programs. **END** is simply a flag to tell the **Toplevel** module to put together a **DEFINE** statement that will define the new function or procedure when it is executed. **TO** and **END** exist simply to provide a convenient way to define new functions and procedures.

† **Exercise 2.5.1.** Define a procedure named **PO.FIRST**, which takes the name of a user-defined function or procedure and displays its definition on three lines. For example, **PO.FIRST "SECOND** should cause the following to be displayed on the text screen:

    Name: **SECOND**
    Input Vars: **OBJECT**
    Body: **[OUTPUT FIRST BF :OBJECT]**

Exercise 2.5.2. Use **DEFINE** to define each of the following functions.

a. A function named **FOURTH** that returns the fourth element of a list.

b. A function named **SWITCH** that takes a word and a list as inputs. It returns a list identical to the input list except that the second element has been replaced by the word input.

Exercise 2.5.3. Define a procedure named **CHANGE.NAME** that takes two words as inputs. The procedure should change the name of the user-defined function or procedure given by the first input to the name given by the second input.

Exercise 2.5.4. Define a procedure named **CHANGE.BODY** that takes a word and a list as inputs. The procedure should change the definition of the operation named by the word input by changing the body to the list input. The name of the operation and the names of the input variables should not change.

2.6 The Editor Module

The **Editor** module has a screen similar to the text screen called the **edit screen**. It differs from the text screen only in that it is larger. It can hold much more than one screenful of text. Only one Logo primitive is associated with the module, but once it has been called, a number of new **editing commands** are made available. After calling up the editor, it takes over interaction with the user from the **Toplevel** module. Returning from the editor returns control to the **Toplevel** procedure.

The **Editor** module is used to modify the definitions of one or more existing user-defined functions and procedures in the Workspace dictionary. It may also be used to add a collection of new ones. The **Editor** module is invoked by using the **edit** procedure:

Edit Procedure

EDIT *object*
EDIT is a procedure that causes the edit screen to be displayed, and the **Editor** module to interact with the user.

The **Editor** can be entered in one of the following three ways:

1. ?EDIT
2. ?EDIT *name*
3. ?EDIT *listofnames*

where *name* is the name of a user-defined function or a procdure and *listofnames* is a list of user-defined functions and procedure names.

In (1) the **Editor** is invoked with an empty edit screen or with its contents identical to the last time the **Editor** was used. The latter will be true if the graphics screen has not been displayed since the last time you used the **Editor**.

In (2), the edit screen will contain the **TO** form of the named function or procedure. The definition will be the one found in the **Workspace** dictionary. If no definition is found with the given name, the edit screen will contain the following line:

TO *name*

The **Editor** is invoked this way to define new functions and procedures. For example, if the procedure **SECOND** has been defined and we enter

?EDIT "SECOND

the edit screen displayed will contain the following text:

TO SECOND :OBJECT
OUTPUT FIRST BF :OBJECT
END

In (3) the edit screen will contain the **TO** forms of all the named functions and procedures. For example, if the functions **SECOND** and **FIRST.TWO** have been defined and we enter

?EDIT [SECOND FIRST.TWO]

the edit screen will contain the following:

TO SECOND :OBJECT
OUTPUT FIRST BF :OBJECT
END

TO FIRST.TWO :OBJECT
OUTPUT SE FIRST :OBJECT FIRST BF :OBJECT
END

When the **Editor** is invoked, you can no longer execute Logo statements, since the **Toplevel** module is no longer in control. The editing commands provided by the editor assist in creating and modifying functions and procedures in the **Workspace** dictionary. To define new ones, just enter their "TO forms" anywhere on the edit screen. They will be added to the workspace dictionary when the editor is exited. Defining operations this way is even more convenient than typing in the "**TO** forms" to the **Toplevel** module because you can go back and modify previously entered lines.

The **Editor** module is described in Figure 2.6, and the commands are described more fully next.

```
┌─────────────────────────────────────────────────────────────┐
│ Editor                                                       │
│                                                              │
│      OBJECTS                                                 │
│         Editscreen:         [[... characters in row 1  ...]  │
│                              [... characters in row 2  ...]  │
│                                      . . .                   │
│                                      . . .                   │
│                                      . . .                   │
│                              [... characters in last row ...]]│
│         Kill Buffer:        [... characters ...]             │
│         Cursor Position:    [CR]                             │
│                                                              │
│      Primitives                                              │
│      EDIT                                                    │
│      EDIT object                                             │
│                                                              │
│      Cursor-Moving Commands                                  │
│         ^A   beginning of line      ^E   end of line         │
│         ^F   forward one space      ^B   back one space      │
│         ^N   next line              ^P   previous line       │
│                                                              │
│      Delete Commands                                         │
│         ^D   delete character       ^H   delete left char    │
│         ^K   moves all characters to                         │
│              the right of the cursor                         │
│              into the "kill buffer"                          │
│                                                              │
│      Special Commands                                        │
│         ^L   center line            ^V   next page           │
│         ^Q   quote next character   ^O   open new line       │
│         ^Y   yank the "kill buffer"                          │
│                                                              │
│      Exit Commands                                           │
│         ^C   install modifications  ^G   no modifications    │
│                                                              │
│      Insert a character                                      │
│         Press a key                                          │
└─────────────────────────────────────────────────────────────┘
```

Figure 2.6 The Editor Module

Exiting the **Editor** and returning control to the **Toplevel** module is done by typing one of the following control characters:

> CONTROL-C returns you to Logo after adding all definitions on the edit screen to the workspace dictionary. If an operation by the same name already exists, it is replaced by the one contained in the edit screen.

> CONTROL-G returns you to Logo without making any additions or modifications to the workspace dictionary.

The editing commands are summarized next. In addition to the four cursor-moving commands seen earlier, the **Editor** has four more:

> CONTROL-A moves the cursor to the beginning of the line.

> CONTROL-E moves the cursor to the end of the line.

> CONTROL-F moves the cursor forward one character.

> CONTROL-B moves the cursor back one character.

> CONTROL-N moves the cursor to the next line.

> CONTROL-P moves the cursor to the previous line.

> CONTROL-V moves the cursor to the next page.

> CONTROL-L moves the line containing the cursor to the center of the screen.

There are three editing commands that are used to delete one or more characters:

> CONTROL-D deletes the character under the cursor.

> CONTROL-H deletes the character to the left of the cursor.

> CONTROL-K moves all characters under and to the right of the cursor into the "kill buffer."

There are two additional commands that provide special features:

> CONTROL-Y copies (yanks) the text contained in the kill buffer into the edit screen starting at the current cursor position.

CONTROL-Q inserts a \ in the text. This tells the reader (**READLIST**) that the next character typed is to be treated as a normal character and that any special meaning attached (such as a delimiting character) should be ignored.

Exercise 2.6.1. Define and execute the following procedure. How can you stop the program from executing over and over again?

```
TO TROUBLE
PRINT [TRY AND STOP ME]
EDIT "TROUBLE
TROUBLE
END
```

2.7 The File System Module

When the power to your computer is turned off, the Logo language, along with the contents of the workspace dictionary, is lost. Loading Logo into your machine again is fast because the Logo system is stored on a diskette.

Typing in all the definitions of operations you need can be very time-consuming, so the Logo language provides the **File System** module, which allows the user to store definitions in the workspace on a diskette. Loading them back into your workspace is as easy as loading Logo.

The **File System** consists of a collection of files. A **file** contains the **TO** forms of user-defined operations, which are stored on a diskette. The contents of a diskette are not lost when the power is turned off.

The primitives associated with the **File System** allow the user to create a file containing the functions and procedures in the workspace dictionary, to add the definitions stored in a file to the workspace dictionary, to find out the names of files on the diskette, and to erase a file from a diskette. The **File System** module is summarized in Figure 2.7 and the primitives that deal with files are described next.

```
┌─────────────────────────────────────────────────────────┐
│ ┌─────────────┐                                         │
│ │ File System │                                         │
│ └─────────────┘                                         │
│                                                         │
│     OBJECTS                                             │
│       Files:    [ file  ...  file  ]                    │
│                                                         │
│     PRIMITIVES                                          │
│                                                         │
│       Procedures                    Functions           │
│         SAVE word                                       │
│         ERASEFILE word                                  │
│         LOAD word                                       │
│         CATALOG                                         │
└─────────────────────────────────────────────────────────┘
```

Figure 2.7 The File System Module

File System Primitives

Procedures

SAVE *word*
SAVE requires the name of a file (a word) as its only input. A file containing the **TO** forms of all dictionary entries is created on the diskette in the computer's disk drive.

LOAD *word*
LOAD requires the name of a file (a word) as its only input. The definitions of the operations contained in the named file will be added to the workspace dictionary. **LOAD** does not change the contents of the named file. If a operation being **LOAD**ed already exists in the dictionary, the definition will be replaced by the one in the file.

ERASEFILE *word*
ERASEFILE requires the name of a file (a word) as its only input. The named file is erased from the diskette.

CATALOG
CATALOG has no inputs. It displays the names of all files on the mounted diskette.

Example 2.7.1.

If the workspace dictionary contains the definitions of **FIRST.TWO** and **SECOND**, then the statement

    SAVE "USEFUL

will create a file named **USEFUL.LOGO** containing the **TO** forms of **FIRST.TWO** and **SECOND**:

```
TO SECOND :OBJECT
OUTPUT FIRST BF :OBJECT
END

TO FIRST.TWO :OBJECT
OUTPUT SE FIRST :OBJECT FIRST BF :OBJECT
END
```

The statement

    LOAD "USEFUL

installs the definitions of **FIRST.TWO** and **SECOND** into the workspace dictionary. Loading the definitions contained in a file back into your workspace does not erase the file on the diskette.

The statement

    CATALOG

displays the names of all the files that currently exist on the mounted diskette.

You can erase the file on a diskette using the **ERASEFILE** statement:

    ERASEFILE "USEFUL

The contents of a file may be changed by erasing the existing file and then issuing another **SAVE** statement using the same file name as its input.

66   The Logo Environment

In order to add one or more definitions to a file, you must first load the current contents of the file into your workspace and then save the old definitions along with the new ones. For example, assume that we have the definition of a function named **THIRD** in the workspace dictionary and that we wish to add it to the **USEFUL.LOGO** file. The following sequence of statements will accomplish the task:

```
?LOAD "USEFUL
?ERASEFILE "USEFUL
?SAVE "USEFUL
```

Exercise 2.7.1. Define a procedure named **UPDATEFILE** that takes a file name as its only input and adds the definitions currently in the workspace dictionary to it. What happens if both the workspace and the named file contain a definition by the same name?

Exercise 2.7.2. Define a procedure named **CREATEFILE** that takes a word as its only input, saves the contents of the workspace in a temporary file, erases all procedures and functions from the workspace, and invokes the editor so the user can add new definitions to the workspace dictionary. When the editor is exited, the workspace is saved in a file whose name is specified by the input word, erases all procedures and functions from the workspace, loads the contents of the temporary file, and then deletes the temporary file.

† **Exercise 2.7.3.** Define a procedure named **EDITFILE** that takes the name of an existing file and a list of function and procedure names as its inputs, saves the contents of the workspace in a temporary file, and erases all procedures and functions from the workspace. It then loads the contents of the file whose name is given by the word input into the workspace, invokes the editor so that the edit screen will contain the definitions of functions and procedures whose names are in the list input, and saves the contents of the workspace in the named file when the editor is exited. The workspace is cleared of all procedures and functions and the temporary file loaded and deleted.

Exercise 2.7.4. Apple Logo allows the user to specify the **slot number** and **drive number** of the disk drive that contains the diskette serving as the **File System**. In addition the user can specify a **volume number,** which identifies the particular diskette that is supposed to be in the indicated disk drive. If the diskette with the indicated volume number is not in the specified disk drive, Apple Logo will inform you so that you do not accidentally save a file on the wrong diskette. In order to handle this, the **File System** must incorporate an additional object, called the **disk-drive object**, and primitives for dealing with this object. The object is a list of three numbers indicating the slot number and drive number that identifies the disk drive and the volume number of the diskette you want in the drive. The primitives that deal with the disk drive object are named **DISK** and **SETDISK**. They display and change the value of the disk-drive object, respectively. Describe the **File System** module that incorporates the diskette-identifying object and the additional primitives that deal with it by redrawing Figure 2.7.

2.8 The Simplifier Module

The **Simplifer** consists of a list of words and lists denoting Logo statements and the **RUN** primitive, which is responsible for simplifying the input expressions for each statement and executing the simplified statements. See **Figure 2.8**.

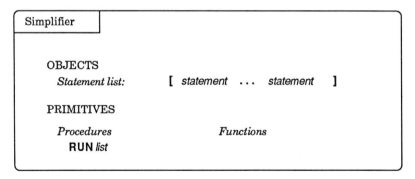

Figure 2.8 The Simplifier Module

> ### The Run Primitive
>
> **RUN** *list*
> **RUN** requires a list as its only input. It views the list as a collection of statements, simplifies the input expressions for the first statement, and calls on the appropriate procedure to perform the specified action. It repeats this action for each of the remaining statements in the list.

The **Simplifier** is the real workhorse of the language. It knows how to perform the primitive operations on words and lists described in Chapter 1. It also performs the simplification of combinations according to the simplification rules described in Chapter 1.

2.9 Memory Layout

Your computer's memory must store the objects described in each of the modules described in this chapter, along with the machine language programs for executing the primitives. A description of the computer's memory is given in Figure 2.9.

Programs for executing procedure and function calls (The Logo system code)
Module objects
Text Screen
Edit Screen
Graphics Screen
Workspace dictionary
Free space

Figure 2.9 Memory Usage

The first section contains all the machine code that handles the execution of the primitive functions and procedures. The next five sections contain the objects used by each of the modules. The three screens and the workspace dictionary are separated out because they occupy significant amounts of memory.

The text screen, edit screen, and graphics screen use large amounts of memory (from 2K to 8K bytes) and are therefore listed separately. They may occupy different areas of memory or they may share the same area to allow more space in the workspace and free space. The sharing of memory is possible, since only one of them may be displayed on the CRT at a time.

In the case of Apple Logo, the edit screen and the graphics screen share the same memory, whereas the text screen is separate. This explains why switching from the graphics screen to the text screen and then back to the graphics screen does not destroy the contents of the graphics screen. On the other hand, switching to the edit screen destroys whatever drawings are on the graphics screen. There are three Logo procedures that control the actual contents of the CRT.

Procedures Affecting the CRT

TEXTSCREEN
TEXTSCREEN causes the CRT to display all 24 lines of the text screen.

FULLSCREEN
FULLSCREEN causes the CRT to display the graphics screen. No text window and hence no **interaction** with Logo is seen on the CRT.

SPLITSCREEN
SPLITSCREEN causes the CRT to display the graphics screen with the most recent four lines of the text screen at the bottom of the CRT.

The section labeled workspace dictionary contains the definitions of all operations defined by the user. The remaining section of memory is called free space and is used as a scratch area during the simplification process.

The boundary between the workspace and free space varies according to how many programs you have added. That is, you start out with no workspace and a lot of free space. As you add definitions, the workspace increases in actual size, whereas the free space decreases in size.

If you keep adding programs to the workspace, eventually you will run out of memory and will not be able to execute your programs. Thus, you will want to keep groups of related definitions in separate files, loading only those that are necessary to run a particular program.

Chapter 3
Extending Logo

3.1 Introduction

This chapter extends the Logo language by introducing the graphics capabilities of Logo, gives more on defining functions and procedures, and expands the ideas of simplification of Logo statements begun in Chapter 1 to include these additions to the language.

When using Logo's graphics facility, two additional modules come into play: the **Turtle** module and the **Graphics Screen** module.

TURTLE		GRAPHICS SCREEN
RIGHT *number* **HEADING**		**CLEAN**
LEFT *number* **TOWARDS** *list*		**CLEARSCREEN**
SETH *number*		**SETBG** *number*
		WRAP
FORWARD *number* **POS**		**FENCE**
BACK *number* **XCOR**		**WINDOW**
SETX *number* **YCOR**		
SETY *number*		
SETPOS *list*		
HOME		
PENDOWN **PEN**		
PENUP **PENCOLOR**		
PENERASE		
SETPC *integer*		
SETPEN *list*		
HIDETURTLE **SHOWNP**		
SHOWTURTLE		

Figure 3.1 Graphics Module

The turtle, which resides on a planar region called the **graphics screen,** is an object with several properties. For example, it has a **position** on the screen as well as a **heading** (the direction it is facing).

Extending Logo

The **graphics screen** is the object on which the turtle runs around. Its properties include its color (this property is called the background color) and the trails left behind by the turtle. The properties of the graphics screen are modified by executing one of the graphics screen primitives or one of several turtle primitives.

Drawing pictures of line segments on the graphics screen is accomplished by entering statements that change the position of the turtle. These statements in turn affect the contents of the graphics screen. That is, most of the changes to the graphics screen (drawing lines) are done indirectly by moving the turtle from one position to another.

The CRT (cathode ray tube) displays the graphics screen when either of the **FULLSCREEN** or **SPLITSCREEN** procedures is executed. Four lines of the **text screen** are displayed at the bottom of the CRT when the **SPLITSCREEN** procedure is executed (see Figure 3.2). In both cases a picture of the turtle is displayed in the center of the graphics screen.

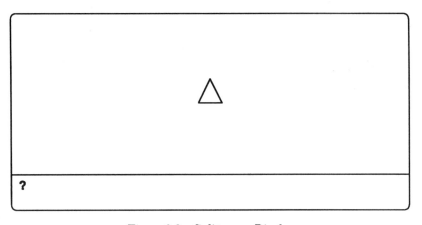

Figure 3.2   Splitscreen Display

We can still use the textscreen as a calculator in **splitscreen mode**. That is, all the things we did in Chapters 1 and 2 can be done in the splitscreen mode. The only difference is that we are only able to see four lines of it. In **fullscreen mode**, the textscreen is not displayed at all.

The entire text screen may be viewed at any time by executing the **TEXTSCREEN** procedure.

3.2 The Turtle

Two properties of the turtle are its **position** and its **heading**. The position is the turtle's location on the graphics screen whereas the heading is the direction the turtle is facing. Initially, the turtle is facing the top of the screen.

The position of the turtle can be changed by executing a **FORWARD** or **BACK** procedure, and the heading can be changed by executing a **RIGHT** or **LEFT** procedure. Each of these procedures requires an input value specifying how far to move or how much to turn.

Another property of the turtle is its **visibility state**. The turtle can be made invisible and then made to reappear at any time by executing the **HIDETURTLE** and **SHOWTURTLE** procedures.

The turtle is also equipped with a pen for drawing. If the pen is down when the turtle moves, the turtle will draw a picture of a line segment. If the pen is up, the turtle will move without drawing. The pen may also be in an erase state, which means that the portions of existing trails that are crossed as the turtle moves will be erased. These three pen states are controlled by the **PENDOWN, PENUP,** and **PENERASE** procedures.

The pen also has an ink color that is controlled by the **SETPC** procedure. It requires an integer describing the ink color as its input value. The following are the colors associated with the integers 0—6.

Pen Colors						
0,6	black	2	green	4	orange
1	white	3	violet	5	blue

The turtle is completely described by the following four properties.

> position
> heading
> visibility
> penstate

We can think of the graphics screen as a surface on which any point can be reached by moving to the right or left a certain distance and then up or down a certain distance. That is, each point on the graphics screen corresponds to an **ordered pair** of numbers. If we **associate** the **ordered number pair** (0, 0) with the center of the screen, we can represent each point on the screen by an ordered pair of numbers that indicates the horizontal and vertical distances from the point associated with (0, 0).

For example, we can think of the ordered pair (20, 30) as representing the point we would arrive at by moving the turtle horizontally to the right 20 steps and vertically up 30 steps from the center of the screen. We can think of the ordered number pair (-20, -30) as representing the point that corresponds to moving the turtle to the left 20 steps and down 30 steps. By giving ordered number pairs to every point on the screen, we can describe the location of the turtle precisely at any time.

The first number in the ordered pair describes how far right or left the turtle has to move from the center of the screen to reach the prescribed point. This number is called the *x*-**coordinate** of the point. The second number in the ordered pair describes how far up or down the turtle must move to get to the point. This number is called the *y*-**coordinate**. The point paired with (0, 0) is called the **origin**.

An ordered pair of numbers can be represented in Logo as a list containing the two numbers. The first number in the list represents the *x*-coordinate and the second represents the *y*-coordinate of the turtle's position. That is, the value of the **position** property is a list of two numbers.

The turtle's heading is a number that is greater than or equal to 0 but less than 360. A 360° heading is the same as a 0° heading. We can ask the turtle to turn any amount we wish, but the direction it winds up facing will be represented as a number in this range. A heading of zero means that the turtle is facing the top edge of the screen. As the turtle rotates to the right, the value of the heading increases until it reaches 360°. Thus, a 90° heading means the turtle is facing the right edge of the screen, and a 180° heading means the turtle is facing the bottom edge of the screen.

The visibility property has one of two values: **TRUE** (the turtle is visible) or **FALSE** (the turtle is invisible).

The three pen positions can be represented by one of the words: **PENUP, PENDOWN,** or **PENERASE,** and the pen color can be represented by a number. Each number corresponds to a particular color. The penstate property of the turtle consists of two independent items: the pen position and the color of its ink. We can represent the penstate property by a list of two elements. The first word is one of the words **PENUP, PENDOWN,** or **PENERASE.** The second element is a nonnegative integer that corresponds to a particular color.

The following are the initial values of the turtle's properties.

position:	[0 0]
heading:	0
visibility:	TRUE
penstate:	[PENDOWN 1]

As in the case of the other modules, there are two types of primitives that can be sent to the turtle. Primitives that modify the properties of the turtle are called **procedures**. They are used to modify the state of the turtle as well as the appearance of the graphics screen. Operations that return a value without modifying the properties of the turtle are called **functions**. They are used to obtain information about the current state of the turtle.

Figure 3.3 describes the **Turtle** module. It consists of an object representing the properties of the turtle and the primitives that deal with these properties. The primitives are grouped according to the property they modify or examine and are described next.

The input expression used with a turtle primitive may be any Logo expression that has the appropriate value for the procedure or function name. Here are some examples:

FD SQRT 50
SETPOS LIST 50 50
RIGHT REMAINDER 980 180
SETH TOWARDS [100 100]

Turtle

OBJECTS

Turtle Properties:

[*position heading visibility penstate*]

PRIMITIVES

Procedures   *Functions*

Heading Property

RIGHT *number*   HEADING
LEFT *number*   TOWARDS *list*
SETH *number*

Position Property

FORWARD *number*   POS
BACK *number*   XCOR
SETX *number*   YCOR
SETY *number*
SETPOS *list*
HOME

Penstate Property

PENDOWN   PEN
PENUP   PENCOLOR
PENERASE
SETPC *integer*
SETPEN *list*

Visibility Property

HIDETURTLE   SHOWNP
SHOWTURTLE

Figure 3.3 The Turtle Module

We need a way to clean up the turtle trails on the graphics screen. The **CLEAN** and **CLEARSCREEN** operations described in the next section provide this utility. They are primitives that belong to the **Graphics Screen** module described in Section 3.5.

Primitives that Deal with the Heading Property

Procedures

RIGHT *number*
RIGHT requires a number as its single input value. The turtle responds by rotating itself to the right through the angle specified by its input value. The name can be abbreviated as **RT**.

LEFT *number*
LEFT requires a number as its only input value. The turtle responds by rotating itself to the left through the angle specified by its input value. The name can be abbreviated as **LT**.

SETHEADING *number*
SETHEADING requires a number as its only input value. The turtle responds by changing its heading to the input value. The name can be abbreviated as **SETH**.

Functions

HEADING
HEADING requires no inputs. A number greater than or equal to 0 and less than 360 is returned. It indicates the turtle's current heading.

TOWARDS *list*
TOWARDS requires a list of two numbers as its only input value. The input value represents a point. The heading that would make the turtle face the indicated point is returned.

Position Property Primitives

Procedures

FORWARD *number*
FORWARD requires a number as its only input value. The turtle responds by moving itself forward the distance specified by the input value. The name can be abbreviated as **FD**.

BACK *number*
BACK requires a number as its only input value. The turtle responds by moving itself backward the specified distance. The name can be abbreviated as **BK**.

SETX *number*
SETX requires a number as its only input value. The turtle responds by moving itself to the point whose x-coordinate is the value of the input and whose y-coordinate is the turtle's current y-coordinate.

SETY *number*
SETY requires a number as its only input value. The turtle responds by moving itself to the point whose y-coordinate is the value of the input and whose x-coordinate is the turtle's current x-coordinate.

SETPOS *list*
SETPOS requires a list of two numbers as its only input value. The turtle will move itself to the point represented by the list and change the value of its position property accordingly.

Functions

POS
POS requires no input values. The turtle responds by returning the current value of its position property.

XCOR
No input values are required by **XCOR**. The turtle responds by returning the x-coordinate of its current position.

YCOR
No input values are required by **YCOR**. The turtle responds by returning the y-coordinate of its current position.

The **HOME** procedure changes two properties of the turtle. It is equivalent to sending **SETPOS [0 0]** followed by **SETH 0**. It is described next:

HOME
HOME requires no inputs. The turtle responds by moving itself to the point [0 0] and changing its heading to 0.

Primitives that Deal with the Pen Property

Procedures

PENDOWN
PENDOWN requires no inputs. The turtle responds by changing its pen position to **PENDOWN** so that the turtle will leave a trail when it moves. It can be abbreviated as **PD**.

PENUP
PENUP requires no inputs. The turtle responds by changing its pen position to **PENUP** so that no trails are drawn when it moves. It can be abbreviated as **PU**.

PENERASE
PENERASE requires no inputs. The turtle responds by changing its pen position to **PENERASE**. The parts of existing trails in the turtle's path are erased when the turtle is moved.

SETPC *integer*
SETPC requires a single input. The turtle responds by changing its pen color to the color represented by the value of its input (one of the integers 0, 1, 2, 3, 4, 5, 6).

SETPEN *list*
SETPEN* requires a single input. The input must be a list containing a pen position (**PENUP, PENDOWN, PENERASE**) and a pen color (a number). The turtle responds by changing the value of its penstate property to the input list.

Functions

PEN
PEN requires no inputs. The turtle responds by returning the value of its penstate property.

PENCOLOR
PENCOLOR requires no inputs. The turtle responds by returning the current value of the pen color.

Primitives that Deal with the Visibility Property Procedures
HIDETURTLE **HIDETURTLE** requires no inputs. The turtle responds by changing the value of its visibility property to **FALSE** and making itself invisible. It can be abbreviated as **HT**. **SHOWTURTLE** **SHOWTURTLE** requires no inputs. The turtle responds by changing the value of its visibility property to **TRUE** and making itself visible. It can be abbreviated as **ST**.

---

* Not in Apple Logo II or IBM Logo.

The Turtle 81

> *Functions*
>
> **SHOWNP**
> **SHOWNP** requires no inputs. The turtle responds by returning the value of its visibility property.

Exercise 3.2.1. Representing a point on a surface by an ordered pair of numbers describing the *x*- and *y*-coordinates of the point is one way it is done in mathematics. Describe an alternative representation of points on a surface.

Exercise 3.2.2. Describe what happens when the following statements are executed.

  a. SETHEADING SQRT 2500
  b. SETHEADING 10 * SQRT 51 + QUOTIENT 40 3
  c. SETHEADING TOWARDS LIST 50 0
  d. SETHEADING TOWARDS POS
  e. SETPOS LIST XCOR + 50 YCOR
  f. SETPOS LIST XCOR YCOR + 50

Exercise 3.2.3. Complete the following chart by placing a check mark in the appropriate column:

Operation	Arity				Function	Procedure
	0	1	2	3 or more		
RIGHT					✓	✓
LEFT						✓
HEADING					✓	✓
SETHEADING						✓
FD						✓
BK						✓
PD						✓
PU						✓
HIDETURTLE					✓	✓
SHOWTURTLE					✓	✓

3.3 The Graphics Screen

The **Graphics Screen** module is described by Figure 3.4. It consists of a representation of a planar region called the **graphics screen** on which we can draw pictures, the properties of the graphics screen, and the primitives that deal with these properties.

The graphics screen is represented by a list, which is described shortly. It has two additional properties: a background color and a boundary type. The primitives that deal with these objects are given on page 83.

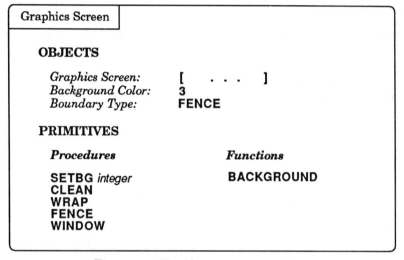

Figure 3.4 *The Graphics Screen Module*

The background colors are identical to those for the turtle's pen. The graphics screen can be defined to have any of three different **boundary types**. The boundary type will cause the turtle to behave differently when it arrives at the display boundary.

A boundary type of **WRAP** affects the movement of the turtle as follows: If the turtle moves to a point on one edge of the graphics screen without changing its heading, it will continue its path from the corresponding point on the opposite side of the screen.

The Graphics Screen **83**

Graphics Screen Primitives

Procedures

CLEAN
CLEAN requires no inputs. All turtle trails are removed from the graphics screen.

CLEARSCREEN
CLEARSCREEN requires no inputs. It causes the turtle to be positioned at the point [0 0] with a heading of zero. It is equivalent to executing the **HOME** procedure and then the **CLEAN** procedure. It can be abbreviated as **CS**.

SETBG *integer*
SETBG requires a number as the value of its single input expression. The color of the graphics screen is changed to the color represented by the input value.

WRAP
WRAP requires no input values. It changes the boundary characteristics of the graphics screen in the following way: Each point on the left edge of the display acts as if it were the same as the point on the right edge of the display that lies on the horizontal line connecting them. Likewise, each point on the top edge of the display is the same as the point on the bottom edge of the display that lies on the vertical line connecting them.

FENCE
FENCE requires no input values. It changes the boundary characteristics of the graphics screen so that the edges of the graphics screen act as rigid barriers for the turtle.

WINDOW
WINDOW requires no input values. It changes the boundary characteristics of the graphics screen so that it is without boundaries. The edges of the display screen simply mark the boundary of the portion of a surface that is centered around the turtle's "home" position.

> **Function**
>
> **BACKGROUND**
> **BACKGROUND** returns a number representing the background color of the graphics screen. It can be abbreviated as **BG**.

When the boundary type is **FENCE** and the turtle is instructed to cross the boundary, the turtle will not move and a turtle out of bounds message will be displayed; no amount of prodding can make the turtle cross this fixed boundary.

If the boundary type is **WINDOW**, the turtle may wander anywhere, but it will not be seen if it is located at a point that does not correspond to a point on the screen. See Figure 3.5.

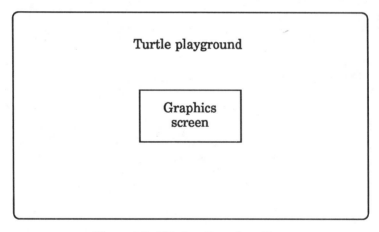

Figure 3.5 Window Boundary Type

This description of the graphics screen is not quite complete because it does not indicate what pictures and drawings it contains. That is, we need a way to represent the pictures that are drawn on the screen. The graphics screen is made up of a collection of small dots placed in rows. See Figure 3.6.

OOOOOOOOOOOOOOOOOOOOOOO ... OOOOOOOOOOOOOOOOOOOOOOO
OOOOOOOOOOOOOOOOOOOOOOO ... OOOOOOOOOOOOOOOOOOOOOOO
OOOOOOOOOOOOOOOOOOOOOOO ... OOOOOOOOOOOOOOOOOOOOOOO
OOOOOOOOOOOOOOOOOOOOOOO ... OOOOOOOOOOOOOOOOOOOOOOO

.
.
.

OOOOOOOOOOOOOOOOOOOOOOO ... OOOOOOOOOOOOOOOOOOOOOOO

Figure 3.6 The Graphics Screen as Rows of Dots

Each dot has a color associated with it. All dots that have the same color as the background color are regarded as the graphics screen's background. To draw a line picture, all dots that are near the points making up the line picture are given the same color as the pen. Since the dots are quite small, we get the visual effect of a line picture drawn on a screen with a given background color. Of course, if the pen color is the same as the background color, we will not see any picture at all.

Each row of dots can be represented as a list of numbers, where the first number represents the color of the first dot in the row, the second number represents the color of the second dot in the row, and so on. The entire graphics screen can be represented as a list of rows. For example, since violet is represented by the number 3 and white is represented by the number 1, Figure 3.7 represents a screen with a violet background containing the picture of a white rectangle in the upper left corner.

```
[ [ 1 1 1 1 1 3 3 3 3 3 3 3 3 3 3 3 3 3 3 3 3 3 3 3 ]
  [ 1 3 3 3 1 3 3 3 3 3 3 3 3 3 3 3 3 3 3 3 3 3 3 3 ]
  [ 1 3 3 3 1 3 3 3 3 3 3 3 3 3 3 3 3 3 3 3 3 3 3 3 ]
  [ 1 1 1 1 1 3 3 3 3 3 3 3 3 3 3 3 3 3 3 3 3 3 3 3 ]
  [ 3 3 3 3 3 3 3 3 3 3 3 3 3 3 3 3 3 3 3 3 3 3 3 3 ]
  [ 3 3 3 3 3 3 3 3 3 3 3 3 3 3 3 3 3 3 3 3 3 3 3 3 ]
  [ 3 3 3 3 3 3 3 3 3 3 3 3 3 3 3 3 3 3 3 3 3 3 3 3 ]
  [ 3 3 3 3 3 3 3 3 3 3 3 3 3 3 3 3 3 3 3 3 3 3 3 3 ]
  [ 3 3 3 3 3 3 3 3 3 3 3 3 3 3 3 3 3 3 3 3 3 3 3 3 ] ]
```

Figure 3.7 White Rectangle on a Violet Background

Exercise 3.3.1. Find out how many rows of dots and how many dots in each row exist on your display screen. The more dots you have to work with, the better the visual effect.

Exercise 3.3.2. Draw the pictures of the following geometric objects:

   a. The largest rectangle that will fit on the screen.
   b. The largest equilateral triangle that will fit on the screen.

3.4 Defining Procedures for the Turtle

New procedures designed to manipulate the turtle can be defined using either the **TO** facility described in Chapter 1 or the **DEFINE** procedure described in Chapter 2. The body of the new procedure is a sequence of procedure calls to which we want the turtle to respond. For example, a procedure that requests the turtle to draw the picture of a square with sides of length 50 could be defined as follows:

```
TO DRAW.SQUARE
FD 50 RT 90
FD 50 RT 90
FD 50 RT 90
FD 50 RT 90
END
```

A **SQUARE** procedure that requires the length of a side as an input value is defined by the following:

```
TO SQUARE :LENGTH
FD :LENGTH RT 90
FD :LENGTH RT 90
FD :LENGTH RT 90
FD :LENGTH RT 90
END
```

If the second form is your choice, you can use the **SQUARE** procedure to draw a square of any size. For example,

```
?SQUARE 100
```

causes the turtle to draw a square whose sides are 100 units long.

Since many times we will want to execute a collection of statements several times, Logo provides a convenient shorthand notation that cuts down on typing:

REPEAT *integer list*
REPEAT is a procedure handled by the **Simplifier**. It requires two input values. The first is a number and the second is a list of statements. The statements in the list are executed the specified number of times.

For example, a square and a rectangle can be drawn by executing the following:

REPEAT 4 [FD 50 RT 90]
REPEAT 2 [FD 50 RT 90 FD 25 RT 90]

The **SQUARE** procedure can be defined more compactly as:

TO SQUARE :LENGTH
REPEAT 4 [FD :LENGTH RT 90]
END

† **Exercise 3.4.1.** For each of the following figures, define a procedure that will cause the turtle to draw the specified picture.

a. Triangle
b. Square
c. Rectangle
d. Hexagon
e. Trapezoid
f. Kite
g. Concave kite
h. Chevron

† **Exercise 3.4.2.** Suppose that the **RIGHT** and **LEFT** procedures were not provided as Logo primitives. Write the Logo definitions of these procedures. Give them the names **RIGHT1** and **LEFT1**. (*Hint:* How would you write a statement that was equivalent to **RIGHT 90** or **LEFT 135**?)

Exercise 3.4.3. Suppose that the **SETX** and **SETY** procedures were not provided as primitives. Write the Logo definitions of these procedures. Give them the names **SETX1** and **SETY1**.

† **Exercise 3.4.4.** Suppose that the **PENUP, PENDOWN, PENERASE,** and **SETPC** procedures were not provided as primitives. Write the Logo definitions of them. Give them the names **PENUP1, PENDOWN1, PENERASE1,** and **SETPC1.***

The **RANDOM** function provides a way to generate random numbers:

> **RANDOM** *integer*
> **RANDOM** is a function that requires a positive integer as its only input value. A random integer greater than or equal to zero and less than the input value is returned.

We can use **RANDOM** together with **REPEAT** to produce random turtle walks. For example,

> **REPEAT 100 [FD RANDOM 10 RT RANDOM 90]**
> **REPEAT 100 [FD RANDOM 10 RT (RANDOM 180)-90]**

causes the turtle to move randomly over the graphics screen.

Exercise 3.4.5. For each of the following, write a Logo function that will generate a random integer n in the range specified:

 a. $1 < n < 6$
 b. $50 < n < 125$
 c. $-45 < n < 45$
 d. $2 < n < 12$

† **Exercise 3.4.6.** Define a procedure that produces a random turtle walk in which the turtle repeats the following two moves 100 times:

---

* SETPEN is required for this exercise. This primitive does not exist in Apple Logo II or IBM Logo.

Defining Procedures for the Turtle 89

1. Moves forward a distance that is at least 5 units but not greater than 15 units.

2. Turns a random number of degrees but never more than 45° to the left or 45° to the right.

Do it once with the pen up and again with the pen down.

Exercise 3.4.7. Define a function that returns the simulated roll of two dice. That is, it should return a list of two numbers, each of which is between 1 and 6 inclusive.

3.5 The Simplification of User-Defined Functions and Procedures

Recall that a user-defined function or procedure has three components: its name, the names of the input variables, and the body:

The three components of a user-defined function or procedure are

1. Its name
2. The names of the input variables (if any)
3. The body

The body consists of a sequence of Logo statements (procedure calls). For example, the body of the following procedure definition is made up of a sequence of eight statements.

```
TO RECTANGLE :WIDTH :LENGTH
FD :WIDTH RT 90
FD :LENGTH RT 90
FD :WIDTH RT 90
FD :LENGTH RT 90
END
```

When the user executes a call to **RECTANGLE**, for example,

    ?RECTANGLE 25 75

the **Simplifier** gets a copy of the definition of **RECTANGLE**:

    [[WIDTH LENGTH] [FD :WIDTH RT 90]
                             [FD :LENGTH RT 90]
                             [FD :WIDTH RT 90]
                             [FD :LENGTH RT 90]]

It then replaces the procedure call by the copy of the body with the actual input values (25 and 75) substituted for the input variables (**:WIDTH** and **:LENGTH**). That is, the simplification proceeds as follows:

    RECTANGLE 25 75

    => FD 25 RT 90
        FD 75 RT 90
        FD 25 RT 90
        FD 75 RT 90

The sequence of statements in the body is then executed in the order they occur.

There is one more situation with which the **Simplifier** must deal. Suppose the user types in the following statement:

    ?RECTANGLE FIRST [25 50 70 100] 60 + 15

The **Simplifier** must first simplify the input expressions and then continue as described. The **Simplifier** needs to perform this step for user-defined functions and procedures as well as for primitive ones.

The simplification of the given statement is described as follows:

    RECTANGLE FIRST [25 50 70 100] 60 + 15

    => RECTANGLE 25 60 + 15
    => RECTANGLE 25 75
    => FD 25 RT 90
       FD 75 RT 90

The Simplification of User-Defined Functions and Procedures   91

    FD 25 RT 90
    FD 75 RT 90

The general rule for simplifying a call to a user-defined function or procedure is described as follows.

> **Simplification of User-Defined Functions and Procedures**
>
> 1. Simplify the input expressions to the named function or procedure.
>
> 2. Replace the simplified call by a copy of the body of the user-defined function or procedure, substituting the first input value for every occurrence of the first input variable, substituting the second input value for every occurrence of the second input variable, and so on.

Exercise 3.5.1. Simplify each of the following statements:

   a. **RECTANGLE (LAST [25 50 75 100]) - 75 3 * SQRT 625.**
   b. **RECTANGLE 100*(SQRT 3)/2 100*(SQRT 2)/2.**
   c. **RECTANGLE LAST [30 40 50] 10*SQRT 64.**
   d. **RECTANGLE (FIRST [40 50 60])+30 70.**
   e. **RECTANGLE LAST BUTFIRST BUTLAST (WORD 3 4 5)**
           BUTLAST (WORD 3 4 5).

3.6 The Simplification of Repeat

A call to the **REPEAT** procedure is simplified by replacing it with the indicated number of copies of the statements in its second input. For example, the simplification of

   REPEAT 4 [RECTANGLE 25 50 RT 90]

is written as follows:

REPEAT 4 [RECTANGLE 25 50 RT 90]

=> **RECTANGLE 25 50 RT 90**
   RECTANGLE 25 50 RT 90
   RECTANGLE 25 50 RT 90
   RECTANGLE 25 50 RT 90

The first statement is a call to a user-defined procedure, so the simplification continues with:

=> **FD 25 RT 90**
   FD 50 RT 90
   FD 25 RT 90
   FD 50 RT 90
   RT 90
   RECTANGLE 25 50 RT 90
   RECTANGLE 25 50 RT 90
   RECTANGLE 25 50 RT 90

Now, the first nine statements are executed, and we are left with:

=> **RECTANGLE 25 50 RT 90**
   RECTANGLE 25 50 RT 90
   RECTANGLE 25 50 RT 90

From here, we continue with:

=> **FD 25 RT 90**
   FD 50 RT 90
   FD 25 RT 90
   FD 50 RT 90
   RT 90
   RECTANGLE 25 50 RT 90
   RECTANGLE 25 50 RT 90

and so on.

The simplification of **REPEAT** is summarized next.

> ## Simplification of Repeat
>
> A call of the form
>
> **REPEAT** *integer statementlist*
>
> is simplified by replacing it with the indicated number of copies of the statements in the statement list.

Exercise 3.6.1. Given the following definition, simplify **FOUR.RECS 10 20**.

```
TO FOUR.RECS :WIDTH :LENGTH
REPEAT 4 [RECTANGLE :WIDTH :LENGTH RT 90]
END
```

Exercise 3.6.2. Define a procedure named **REGULAR.POLYGON** that requires two input values: the number of sides of the polygon and the measure of each side. When called, the procedure will cause the turtle to draw a picture of a regular polygon on the screen. For example,

REGULAR.POLYGON 3 50

will result in a picture of an equilateral triangle with sides of length 50, and

REGULAR.POLYGON 4 100

will result in a picture of a square of side measure 100.

3.7 Recursive Procedures

Any user-defined function or procedure can call on any other user-defined function or procedure, including itself. That is, we can define procedures that cause another copy of itself to be executed. For example, consider the following definition:

```
TO DOSQUARE :DIST
FD :DIST
RT 90
DOSQUARE :DIST
END
```

We can analyze the execution of a call to **DOSQUARE** using the simplification process. For example, the simplification of **DOSQUARE 50** is described as follows:

DOSQUARE 50

```
=> FD 50
   RT 90
   DOSQUARE 50
```

After executing **FD 50** and **RT 90**, the **Simplifier** is left with

```
=> DOSQUARE 50
```

which simplifies to:

```
=> FD 50
   RT 90
   DOSQUARE 50
```

Again, **FD 50** and **RT 90** are executed and we continue with

```
=> DOSQUARE 50
```

```
=> FD 50
   RT 90
   DOSQUARE 50
```

and so on forever.

Procedures that call themselves are called **recursive** procedures. Executing **DOSQUARE 50** causes the turtle to trace the picture of a square over and over again. The computation will not end until someone cuts the power to your computer (or types CONTROL-G).

Exercise 3.7.1. Describe the execution of each of the following by showing all steps in the simplification.

    MYSTERY 50 10.
    MYSTERY 50 180

where **MYSTERY** is defined as follows:

    TO MYSTERY :SIZE :ANGLE
    RECTANGLE :SIZE :SIZE
    RT :ANGLE
    MYSTERY :SIZE :ANGLE
    END

Exercise 3.7.2. Describe what happens when **BEEP** is executed, where **BEEP** is defined as follows:

    TO BEEP
    PRINT CHAR 7
    BEEP
    END

Exercise 3.7.3. Describe what happens when **FLASH 0** is executed, where **FLASH** is as defined. Describe the execution of **FLASH 10** using substitution and simplification.

    TO FLASH :N
    SETBG REMAINDER :N 6
    FLASH :N + 1
    END

If we change the definition of **DOSQUARE** by allowing the angle to become an input, we can generate other geometric figures. Instead of redefining **DOSQUARE**, let's define a new operation called **POLY** as follows:

    TO POLY :DISTANCE :ANGLE
    FD :DISTANCE
    RT :ANGLE
    POLY :DISTANCE :ANGLE
    END

96 Extending Logo

Here are some other variations on the same theme:

```
TO GROW :DISTANCE :ANGLE
FD :DISTANCE
RT :ANGLE
GROW :DISTANCE + 2 :ANGLE
END

TO DESIGN :DISTANCE :ANGLE
FD :DISTANCE RT :ANGLE
FD :DISTANCE RT :ANGLE
DESIGN :DISTANCE :ANGLE + 10
END
```

Exercise 3.7.4. Execute the **POLY** procedure using each of the following input values for **:ANGLE** and your own choice for **:DISTANCE**:

60 72 144 135 108 10

Exercise 3.7.5. Using the following values for **:ANGLE**, execute calls to **GROW** and **DESIGN**. Use your own choice for **:DISTANCE**.

90 95 120 117

Exercise 3.7.6. Using the following definition for **SPI**, execute calls to **SPI** using each of the following pairs of values for **:ANGLE** and **:INC**. Use the same value for **:C** each time.

```
TO SPI :DISTANCE :ANGLE :INC
FD :DISTANCE
RT :ANGLE
SPI :DISTANCE :ANGLE + :INC :INC
END
```

| :ANGLE | 2 | 40 | 0 |
| :INC | 20 | 30 | 7 |

Exercise 3.7.7. A *golden rectangle* is one whose length is 1.61803 times its width.

   a. Define a Logo procedure requiring a single input value to draw golden rectangle pictures of any desired size.

   b. Define a Logo procedure that draws a spiral of golden rectangle pictures.

3.8 Procedures vs. Functions

The previous sections were concerned with defining procedures that cause the turtle to draw pictures, the graphics screen to flash its background in different colors, and the computer to beep.

Procedures do not return values. They are executed for the **side effects** they cause. A side effect is a change in state that is a result of executing one or more statements. The side effect caused by executing an **FD** or **SETPOS** statement is the modification of the values of the properties of the turtle as well as the graphics screen. The state of the textscreen is modified by **SHOW** or **POTS**.

Functions are designed to return values that can be used any way we desire. They do not cause any changes (side effects) to the turtle, the graphics screen, the text screen, the workspace dictionary, or to anything else in the Logo world. For example, **POS** returns the current value of the turtle's position, and **2 + 3** returns the number **5**. **POS** and **+** are functions.

"Writing a program" means to define a function or a procedure. If the purpose of the program is to draw a picture on the graphics screen, the program should be a procedure. If the purpose is to compute a value, the program should be a function.

If the programming task is to compute a value, we are often tempted to define a procedure to do the task. For example, if we want to write a program that computes the square of a number, we might come up with the following definition:

```
TO SHOW.SQ :X
SHOW :X * :X
END
```

SHOW.SQ is a procedure. It causes a side effect. The value computed is displayed on the text screen. It does not return a value that can be used as an input to some other operation.

If our next programming task is to compute the sum of the squares of two numbers, the **SHOW.SQ** procedure would be of no value whatsoever, because it does not return a value. If we had defined our squaring program as a function,

```
TO SQ :X
OUTPUT (:X * :X)
END
```

we could then use it in defining the sum-of-two-squares program. That is, we could use the expression **(SQ :X) + (SQ :Y)**, where **:X** and **:Y** are the input variables to the sum of two squares program. The sum-of-two-squares program should also be written as a function because we might be able to use it in yet another program (see Exercises 3.8.2 and 3.8.3):

```
TO SUM.SQ :X :Y
OUTPUT (SQ :X) + (SQ :Y)
END
```

Rather than **SHOW**ing the computed value, we want to **OUTPUT** it. Programs that **SHOW** values are of no use in the computation of something more complex. You cannot use the value displayed as an input to another operation.

The programmer never knows when a program to compute a value might be useful in a later programming task. Thus, you should always write such a program as a function. You can always display the result using the **SHOW** procedure at the Logo's top level:

```
?SHOW SQ 8
64
?SHOW SUM.SQ 3 4
25
?
```

The description of **OUTPUT** and its simplification are as follows.

> **OUTPUT** is a unary procedure that can be used only in the body of a function. It causes the function to return the value of the expression following the word **OUTPUT**.
>
> A statement of the form **OUTPUT** *expression* is **simplified** by replacing the entire sequence of statements in the body of the function with *expression*.

Using the rule for simplification, you can simplify **SUM.SQ FIRST [4 5 6] 3** as follows:

SUM.SQ FIRST [4 5 6] 3

```
=> SUM.SQ 4 3
=> OUTPUT (SQ 4) + (SQ 3)
=> (SQ 4) + (SQ 3)
=> (OUTPUT 4 * 4) + (SQ 3)
=> (4 * 4) + (SQ 3)
=> 16 + (SQ 3)
=> 16 + OUTPUT 3 * 3
=> 16 + 3 * 3
=> 16 + 9
=> 25
```

The next example shows how to simplify a user-defined function that contains an output statement as the first of a sequence of statements. If **WHERE** is defined by

```
TO WHERE
OP POS
FD 50
OP POS
END
```

and the position of the turtle is given by **[100 100]** then the simplification of **WHERE** is performed as follows:

```
WHERE
=> OP POS
   FD 50
   OP POS
=> POS
=> [100 100]
```

The **FD 50** and the second **OP POS** statements are never executed.

We might argue that we have the best of both worlds if we define our squaring and sum-of-squares programs as follows:

```
TO SQ1 :X
SHOW (:X * :X)
OUTPUT (:X * :X)
END


TO SUM.SQ1 :X :Y
SHOW (SQ1 :X) + (SQ1 :Y)
OUTPUT (SQ1 :X) + (SQ1 :Y)
END
```

With these definitions we really have something ugly:

```
?SHOW SQ1 8
64
64
?SHOW SUM.SQ1 3 4
9
16
25
9
16
25
?
```

The lesson to learn is *never* to have a function display a value. Programs such as **SQ1** and **SUM.SQ1** are called **pseudofunctions**. They return values as well as display them. Since we usually do not always know when such a function is called, it appears that random numbers occur on the screen. They are more confusing than helpful when you are trying to understand what a program is doing.

Exercise 3.8.1. Define a function named **CUBE** that takes a number as its single input. **CUBE** returns the cube of its input value. For example, **CUBE 3** returns **27**.

Exercise 3.8.2. Define a function named **DISTANCE** that computes the distance between two points in the plane. The function has two input variables, one for each point. Each point is assumed to be represented by a list containing its *x*- and *y*-coordinates. For example, **DISTANCE [3 4] [0 0]** returns **5**.

Exercise 3.8.3. Define a Logo function named **HOW.FAR** that returns a number indicating how far the turtle is from its current position to its home position.

Exercise 3.8.4. Define a function named **FIRST.TO.LAST** that takes a word as its only input. **FIRST.TO.LAST** returns a word constructed by moving the first character of its input to the end of the word.

Exercise 3.8.5. Define a function named **SECOND.TWO** that takes a word as its only input. **SECOND.TWO** returns a word consisting of the second and third characters of its input. For example, **SECOND.TWO "GLOB** returns the word **LO**.

Exercise 3.8.6. Define a function named **SWITCH** that takes a list as its single input. **SWITCH** returns a list whose elements are the same as the elements of its input list except that the first and last elements are reversed. For example, **SWITCH [A B C D]** returns the list **[D B C A]**.

Exercise 3.8.7. Define a Logo function named **BLEND.TWO** that takes a list of words as its single input. **BLEND.TWO** returns a word formed by combining the first two elements of its input into a single word. For example, **BLEND.TWO [BIG BLUE BOX]** returns the word **BIGBLUE**.

Exercise 3.8.8. Define a Logo function named **FIRST.TWOP** that has two inputs, both of which are words. **FIRST.TWOP** returns the word **TRUE** if the first two characters of the first input are equal to the first two characters of the second input. It returns **FALSE** otherwise. For example, **FIRST.TWOP "EXAMPLE "AXE** returns **FALSE**, whereas **FIRST.TWOP "EXAMPLE "EXEMPT** returns **TRUE**.

Exercise 3.8.9. Define a Logo function named **DIV3** that needs a number as its single input. **DIV3** returns the word **TRUE** if its input is divisible by 3 and **FALSE** otherwise. For example, **DIV3 12** returns **TRUE**, whereas **DIV3 17** returns **FALSE**.

3.9 Program-Defining Programs

You may use either the **DEFINE** procedure or the **TO** form to add a new definition into the workspace dictionary. The **TO** form, however, can only be used when you are interacting with Logo at the top level (when entering statements after the ? prompt). The **DEFINE** procedure can be called from the body of a procedure as well as from the top level.

As an example of using **DEFINE** in the body of another procedure consider the following definition:

```
TO DEFINEPOLY :N
DEFINE
   (WORD "POLY :N)
   LIST [SIZE]
        (LIST "REPEAT :N (SE [FD :SIZE RT] 360/:N))
END
```

DEFINEPOLY requires a number as its only input value. To see what is going on here, we trace the behavior of the **Simplifier** when the following statement is entered:

   ?DEFINEPOLY 5.

After substituting the number 5 for every occurrence of :N, we have the following:

DEFINEPOLY 5

(1)
**=> DEFINE (WORD "POLY 5)
     LIST [SIZE] (LIST "REPEAT
                        5
                        (SE [FD :SIZE RT] 360/5))**

(2)
**=> DEFINE "POLY5
     LIST [SIZE] (LIST "REPEAT 5 [FD :SIZE RT 72])**

(3)
=> DEFINE "POLY5 [[SIZE] [REPEAT 5 [FD :SIZE RT 72]]]

Here, (2) is obtained by simplifying the first input expression and (3) is obtained by simplifying the second input expression.

Next, the call to **DEFINE** is executed, adding a new definition to the workspace dictionary. The name of the new procedure is **POLY5**. It requires one input value (a number), and the body of **POLY5** is given by the list **[REPEAT 5 [FD :SIZE RT 72]]**. To verify, enter

?PO "POLY5

The following definition is displayed:

**TO POLY5 :SIZE
REPEAT 5 [FD :SIZE RT 72]
END**

That is, the execution of **DEFINEPOLY 5** causes this definition to be added to the workspace. Try it out by entering

?POLY5 20

DEFINEPOLY is a procedure that defines other procedures. The procedures it defines draw pictures of polygons with a particular number of sides. Executing **DEFINEPOLY 8** adds a procedure named **POLY8**, which can then be used to draw octagons of any desired size.

Chapter 4
The Ultimate In Computational Power

4.1 Introduction

Our description of Logo so far is still that of a calculator—although programmable. We have the ability to define new operations for the calculator, but it is still rather limited. An example of a function that we cannot define with the features previously discussed is a function that returns the length of a list. There are many more.

In this chapter we introduce a primitive named **IF**, which will give us the capability to define functions and procedures that will execute one collection of statements if a certain condition is true and to execute a different collection of statements if the condition is false. This capability will give us a big jump in power. In fact, the addition of **IF** gives Logo all the computational power that any programming language can have.

4.2 Recognizers

A **recognizer** is a function that tells us whether or not a certain condition is **true** or **false**. That is, it is a function that returns a truth value. Three such functions are =, <, and >. In this section we introduce additional primitive recognizers and define several more.

Examples of other conditions we might want to recognize are given in the following statements:

1. The number n is less than or equal to the number m.
2. The number n is greater than or equal to the number m.
3. The number n is greater than m and less than k.

We could add operations to our language to handle each of these, but this is a shortsighted solution, since we will always find the need to add more. We follow the lead in the last chapter by defining new operations in terms of the ones that exist.

The first two examples ask if either of two conditions is true, whereas the third asks if two conditions are true at the same time. Extending this idea slightly and generalizing, we decide that we want to be able to answer questions like the following:

4. Are any of a collection of conditions true?
5. Are all the conditions in a collection true?

The functions named **OR** and **AND** give us the desired capability. They are described below along with a function named **NOT**. They form the **logical operations** of Logo.

Logical Operations

OR *exp ... exp*
OR requires two or more input expressions that evaluate to *true* or *false*. **TRUE** is returned if any of the input expressions are **TRUE**. Otherwise, **FALSE** is returned.

AND *exp ... exp*
AND requires two or more input expressions that are evaluated as *true* or *false*. **TRUE** is returned if all expressions are **TRUE**. Otherwise, **FALSE** is returned.

NOT *exp*
NOT is a unary operation. If the input expression is evaluated as *true*, **FALSE** is returned. Otherwise, **TRUE** is returned.

Since **OR** and **AND** may have an arbitrary number of input values, parentheses must surround the entire expression when more than two are specified.

We can define a recognizer that checks to see if one number is greater than or equal to another number by

```
TO GEP :N1 :N2
OP OR :N1 > :N2  :N1 = :N2
END
```

and a recognizer named **BETP** that checks to see if a number n lies between two other numbers by

```
TO BETP :N :LIST
OP AND (FIRST :LIST) < :N  :N < (LAST :LIST)
END
```

where it is assumed that the smaller number is the first element of **:LIST** and the larger is the second.

These recognizers have names that end with the letter **P**. This is just a convention that is often used to indicate that the function is a recognizer, that is, one that returns either the word **TRUE** or the word **FALSE**. Such functions are called **predicates**.

Sometimes it is important to know when an input is a particular kind of object. For example, the Logo language itself must determine what kind of input value (a word or a list) is specified with operations such as **FIRST** or **BUTFIRST**.

Extracting the first character of a word is different from extracting the first element of a list.

Logo provides primitive functions that may be used to recognize the occurrence or a word, a list, or a word that represents a number. Another useful recognizer, **EMPTYP**, is used to recognize the occurrence of an empty list or word.

Recognizers
WORDP *object* **WORDP** is a unary function. If the value of the single input is a word, then the word **TRUE** is returned. Otherwise, **FALSE** is returned.

> **LISTP** *object*
> **LISTP** is a unary function. If the value of its single input is a list, then the word **TRUE** is returned. Otherwise, **FALSE** is returned.
>
> **NUMBERP** *object*
> **NUMBERP** is a unary function. If the value of its single input is the base 10 name of a number, then the word **TRUE** is returned. Otherwise, **FALSE** is returned.
>
> **EMPTYP** *object*
> **EMPTYP** is a unary function. If the value of its single input is the empty list (a list with zero elements), **TRUE** is returned. Otherwise, **FALSE** is returned.

Example 4.3.1.
```
?SHOW WORDP "ROBOT
TRUE
?SHOW WORDP [A B C]
FALSE
?SHOW WORDP 374
TRUE
?SHOW LISTP BUTFIRST "ROBOT
FALSE
?SHOW LISTP BUTFIRST [A B C]
TRUE
?SHOW NUMBERP BUTFIRST "ROBOT
FALSE
?SHOW NUMBERP 374
TRUE
?SHOW NUMBERP FIRST [123.45 RIPE ORANGES]
TRUE
?SHOW EMPTYP BUTFIRST [TABLE]
TRUE
?SHOW EMPTYP BUTFIRST "CHAIR
FALSE
```

We can define new recognizers using these primitive functions. For example, here is the definition of a recognizer that tells us whether or not the first and last character of a word are the same character:

```
TO FIRST.EQUALS.LASTP :OBJECT
OP (FIRST :OBJECT) = LAST :OBJECT
END
```

Note that this definition doesn't distinguish between word or list inputs. It returns **TRUE** if its input is a word and the first and last characters are the same. It also returns **TRUE** if its input is a list and the first and last elements match.

Exercise 4.2.1. Define a recognizer named **LEP**, that requires two numbers as input values. It returns **TRUE** if the first input is less than or equal to the second and **FALSE** otherwise.

Exercise 4.2.2. Define a prefix recognizer named **EQP**, which requires two numbers as input values. **EQP** returns **TRUE** if the input values are equal and **FALSE** otherwise.

Exercise 4.2.3. Define a prefix recognizer named **LTP** that requires two numbers as input values. It returns **TRUE** if the first input value is less than the second and **FALSE** otherwise.

Exercise 4.2.4. Define a prefix recognizer named **GTP**, which requires two numbers as input values. It returns **TRUE** if the first input value is greater than the second and **FALSE** otherwise.

The prefix functions defined in these exercises can be used in place of their infix cousins. This is useful because we then will have a complete set of prefix operations for comparing numerical values:

```
LTP
GTP
EQP
LEP
GEP
```

Exercise 4.2.5. The **BETP** function can be used to check to see if a number lies between two other numbers but is not equal to either one. Define functions named **EQBETP**, **BETEQP**, and **EQBETEQP** to check to see if a number lies between two numbers or is equal to the first, lies between two numbers or is equal to the second, and lies between two numbers or is equal to either one. For example,

```
BETP 3 [2 3] returns FALSE
BETP 2 [2 3] returns FALSE

EQBETP 3 [2 3] returns FALSE
EQBETP 2 [2 3] returns TRUE

BETEQP 3 [2 3] returns TRUE
BETEQP 2 [2 3] returns FALSE

EQBETEQP 3 [2 3] returns TRUE
EQBETEQP 2 [2 3] returns TRUE
```

4.3 The IF Procedure

The **IF** procedure recognizes whether a certain condition is true or false and executes one collection of statements in the first case and another collection of statements in the other.

The IF Procedure

IF *condition list*
IF *condition list1 list2*
IF requires either two or three input values. The first is an expression that is evaluated as **TRUE** or **FALSE**. The second and (optional) third input values must have lists as their values. The elements of each list represent a sequence of Logo statements.

> If the value of *condition* is the word **TRUE**, then the statements in *list1* are executed.
>
> If the value of *condition* is **FALSE**, and *list2* is present, then the statements in *list2* are executed.
>
> If the value of *condition* is **FALSE** and *list2* is not present, then the statement following the **IF** statement is executed.

Example 4.3.1 Define a Logo function named **ABS**, which computes the absolute value of a number.

Algorithm The absolute value of a number is defined to be the number itself if the number is greater than or equal to zero and the negative of the number if the number is less than zero. That is,

$$abs(num) = \begin{cases} num & \text{if } num \geq 0 \\ -num & \text{if } num < 0 \end{cases}$$

where *num* is the number whose absolute value we want.

Logo Definition Defining the absolute value function in Logo is just a straightforward translation from the language of mathematics.

```
TO ABS :NUM
IF GEP :NUM 0 [OP :NUM] [OP -:NUM]
END
```

The body of **ABS** consists of one statement and must be entered on a single line. We use the convention that whenever a line is indented beyond the first letter of the line above, it should be regarded as a continuation of the line above.

Here is another way to define the absolute value function:

```
TO ABS :NUM
IF LTP :NUM 0  [OP -:NUM] [OP :NUM]
END
```

The body of this definition uses the *less-than* predicate, **LTP**, to do the test. Using this test requires the list inputs to be reversed.

We could use the **IF** without the second list as an input to **IF** by writing:

```
TO ABS :NUM
IF LTP :NUM 0 [OP -:NUM]
OP :NUM
END
```

Frequently, the statements to execute next depend on more than one condition being true, or they might depend on one of several conditions being true. For example, consider the following function definition:

$$step(num) = \begin{cases} 1 & \text{if } 0 \leq num \leq 2 \\ 0 & \text{otherwise} \end{cases}$$

The definition of **STEP** can be written as

```
TO STEP :NUM
IF AND GEP :NUM 0 LEP :NUM 2 [OP 1] [OP 0]
END
```

or by

```
TO STEP :NUM
IF OR GTP :NUM 2 LTP :NUM 0 [OP 0] [OP 1]
END
```

Example 4.3.2 illustrates a case that involves words.

Example 4.3.2 Define a unary function named **STRIP** that requires a word containing at least two characters as the value of its input. If the first and last characters of the input are identical to the character input, **STRIP** returns a word formed by removing the first and last characters of its input. Otherwise, it returns a word that is the same as its input.

112 *The Ultimate in Computational Power*

Algorithm The algorithm is straightforward:

If the first and last character of the word input are the same, return the word with the first and last characters of the input removed. Otherwise, return the word itself.

Logo Definition Recalling the definition of **FIRST.EQUALS.LAST** in Section **4.2**, we arrive at the following definition:

```
TO STRIP :WORD
IF FIRST.EQUALS.LAST :WORD
   [OP BUTFIRST BUTLAST :WORD]
   [OP :WORD]
END
```

Exercise 4.3.1. Define **STEP** using **EQBETEQP**.

Exercise 4.3.2. Define a function named **INTEGERP** that takes a number as its single input. It returns the word **TRUE** if the input is an integer and the word **FALSE** if the input is not an integer.

Exercise 4.3.3. Define a unary function named **ODD.OR.EVEN**. It returns the word **ODD** if the input is an odd integer, the word **EVEN** if its input is an even integer, and the word **NEITHER** if the number is not an integer.

4.4 Counting Recursion

The inclusion of **IF** does not give us much in the way of additional computational power until it is coupled with recursion. Together they give us all the computational power any programming language can have. In this section, several examples of a special form of recursion called *counting recursion* are examined. You are advised to study them closely. A complete understanding of how to use counting recursion and the techniques described in the remaining sections of this chapter will provide you with a powerful tool for describing complex problems in a clear and concise fashion.

Example 4.4.1 Define a Logo function named **power** that computes the nth power of a number where n is a nonnegative integer.

> **Algorithm 1**   The nth power of a number is formed by finding the product of n "copies" of the number.

This algorithm is not very helpful for computational purposes. It says to define a logo expression of the form :X * :X ... * :X with the appropriate number of :X's, evaluate it, and return the result. A more natural algorithm comes directly from mathematics.

> **Algorithm 2**   The nth power of a number is computed by multiplying the number by the $(n-1)$st power of the number. Of course, the result is 1 if n is zero. Thus, we have:

$$x^n = \begin{cases} 1 & \text{if } n = 0 \\ x * x^{n-1} & \text{if } n > 0 \end{cases}$$

> **Logo Definition**   The function requires two inputs: a number x and a nonnegative integer. The Logo definition is a straightforward translation of Algorithm 2:

```
TO POWER :X :N
IF :N = 0 [OP 1] [OP :X * POWER :X :N-1]
END
```

POWER is an example of a recursively defined function that will terminate provided that the value of :N in the initial call is positive. If the initial value of :N is negative, :N will keep getting smaller and smaller in the negative direction with each recursive call to itself, and the computation will never stop. Actually, Logo will eventually run out of free space when executing something like

```
POWER 2 -3
```

because it keeps generating more and more new computations to perform. When it does run out of free space, it will complain.

The definition of **POWER** illustrates **counting recursion**. The value of one of its inputs serves as a counter for the number of times the function calls itself. It can be characterized by the two components described next.

Counting Recursion

Termination Case The value of the function is known immediately for a small value (usually, zero or one) of the counter input.

General Case The value of the function can be computed by performing an operation that uses the value obtained by calling itself with the counter decreased by one.

In the **POWER** example, the terminating case occurs when the value of n is zero. The general case tells us what to do if the condition for the terminating case is not met. It consists of returning the product of **X** with **X** raised to the $(n - 1)$st power.

We can use the substitution and simplification model of computation to see how this works on a specific example, computing 4 to the third power. We begin by substituting **4** for **:X** and **3** for **:N** in the body of **POWER**.

```
POWER 4 3
=> IF 3 = 0 [OP 1] [OP 4 * POWER 4 3-1]
```

Now we must simplify the **IF** expression. The rule used to simplify an **IF** statement is described next.

IF *condition list1 list2* is simplified as follows:

1. Simplify *condition*.

2. If the result in (1) is **TRUE**, then replace the entire **IF** statement by the statements in *list1*.

> 3. If the result in (1) is **FALSE** and *list2* is present, then replace the entire **IF** statement by the statements in *list2*.
>
> 4. If the result in (1) is **FALSE** and *list2* is not present, then the **IF** statement is discarded.

We can now simplify **POWER 4 3**:

```
POWER 4 3    => IF 3 = 0 [OP 1] [OP 4 * POWER 4 3-1]
             => OP 4 * POWER 4 3-1
             => 4 * POWER 4 3-1
             => 4 * POWER 4 2
             => 4 * (IF 2 = 0 [OP 1] [OP 4 * POWER 4 2-1])
             => 4 * (OP 4 * POWER 4 2-1)
             => 4 * (4 * POWER 4 2-1)
             => 4 * (4 * POWER 4 1)
             => 4 * (4 * (IF 1 = 0 [OP 1] [OP 4 * POWER 4 1-1]))
             => 4 * (4 * (OP 4 * POWER 4 1-1))
             => 4 * (4 * (4 * POWER 4 1-1))
             => 4 * (4 * (4 * POWER 4 0))
             => 4 * (4 * (4 * (IF 0 = 0 [OP 1] [OP 4 * POWER 4 0-1])))
             => 4 * (4 * (4 * OP 1))
             => 4 * (4 * (4 * 1))
             => 4 * (4 * 4)
             => 4 * 16
             => 64
```

Another example of counting recursion involves finding the nth element of a list.

Example 4.4.2 Define a function named **NTH**, which has a nonempty list as its first input and a positive integer n as its second input. The nth element of a list is returned by *nth*. For example,

> NTH [A B C D] 3 returns C

Algorithm How does this example use counting recursion? The termination case here occurs when n is 1. If $n = 1$, the result is the first element of the list. Otherwise, the result is the same as finding the $(n - 1)$st element of the rest of the list. That is,

$$nth\,(l, n) = \begin{cases} \text{first } l \text{ if } n = 1 \\ nth\,(bf\ l, n-1) \text{ if } n > 1 \end{cases}$$

Logo Definition*

```
TO NTH :LIST :N
IF (:N = 1)
    [OP FIRST :LIST]
    [OP NTH BUTFIRST :LIST :N-1]
END
```

Exercise 4.4.1. Describe the evaluation of **NTH [A B C D] 3** using substitution and simplification.

Exercise 4.4.2. Define the following functions.

a. A function named **FACT** that takes a number n as its only input and returns n-factorial, where n-factorial is denoted by $n!$ and is defined as follows:

$n! = n(n-1)(n-2)...1$

or, more compactly, by: $n! = n(n-1)!$.

b. A Logo function named **NTHCHAR** that takes a word and a number n as its input values and returns the nth character of the word.

4.5 List Recursion

Another form of recursion is encountered in the following problem.

Example 4.5.1 Define a function that will return the average of a collection of numbers.

---

* The Logo primitive named ITEM is equivalent to NTH.

Analysis The algorithm to solve this problem is very familiar. Simply add all the numbers and divide by the size of the collection.

If we are to define a Logo function to perform this task, we must decide how to represent the collection of numbers. The way to represent a collection of objects in Logo is by using a list. Our Logo function—let's name it **AVERAGE**—will have a list of numbers as its single input value.

Logo Definition

```
TO AVERAGE :LIST
OP (ADD :LIST) / LENGTH :LIST
END
```

There are two problems with this definition. The first is that we do not have functions named **ADD** and **LENGTH** to do the job for us, and the second is that if the length of the list is zero, we will be dividing by zero. The second problem is solved with a slight modification of the above definition:

```
TO AVERAGE :LIST
IF EMPTYP :LIST
  [OP 0]
  [OP (ADD :LIST) / LENGTH :LIST]
END
```

The first problem is solved by defining an operation named **ADD**, which computes a sum, and an operation named **LENGTH**, which computes the number of elements in the list.

Subproblem Define a function named **ADD**, which has a list of numbers as the value of its input expression. **ADD** returns the sum of those numbers.

More Analysis If we knew how many elements were in the list we could use counting recursion. However, we do know how to add a collection of zero numbers. The sum is zero. Now, if there is at least one number in the list, add the first number in the list to the sum of the rest of the numbers in the list. That is, we have:

A **termination case**: The answer is zero if the list is empty.

A general case: Add the first element of the list to the sum of the rest of the elements in the list.

The Logo definition of **ADD** is:

```
TO ADD :LIST
IF EMPTYP :LIST
  [OP 0]
  [OP (FIRST :LIST) + ADD BF :LIST]
END
```

Using the substitution and simplification model of computing, we can simplify **ADD [3 5 7]** as follows:

```
ADD [3 5 7]
=> IF EMPTYP [3 5 7]
     [OP 0]
     [OP (FIRST [3 5 7]) + ADD BF [3 5 7]]
=> (FIRST [3 5 7]) + ADD BF [3 5 7]
=> 3 + ADD [5 7]
=> 3 + (IF EMPTYP [5 7]
          [OP 0]
          [OP (FIRST [5 7]) + ADD BF [5 7]])
=> 3 + ((FIRST [5 7]) + ADD BF [5 7])
=> 3 + (5 + ADD [7])
=> 3 + (5 + (IF EMPTYP [7]
               [OP 0]
               [OP (FIRST [7]) + ADD BF [7]]))
=> 3 + (5 + ((FIRST [7]) + ADD BF [7]))
=> 3 + (5 + (7 + ADD [ ]))
=> 3 + (5 + (7 + (IF EMPTYP [ ]
                    [OP 0]
                    [OP (FIRST [ ])
                          + ADD BF [ ]])))
=> 3 + (5 + (7 + 0))
=> 3 + (5 + 7)
=> 3 + 12
=> 15
```

It is important to remember how an **IF** statement is simplified. In the simplification of

```
IF EMPTYP [ ]
   [OP 0]
   [OP (FIRST [ ]) + ADD BF [ ]]
```

we replace the entire expression with one of the two alternatives (either **OP 0** or **OP (FIRST [ ]) + ADD BF [ ]**) after determining the value of the condition. We do not simplify the alternative computations until it has been determined which to use!

If we tried to simplify **FIRST [ ]** or **BF [ ]** before simplifying the **IF** expression, we would get an undefined result (an error message). The rule for simplifying an **IF** expression guarantees that we will never be faced with simplifying **BF [ ]** in Example 4.5.1.

ADD illustrates a second type of recursion, called **list recursion**. List recursion is appropriate when the definition of a function has a list as an input, when we do not know how many elements are in the list, and when we need to perform some computation for each element in the list.

In **ADD**, we select the first element and add it to the sum of the rest of the elements. That is, we do an addition for each element in the list. In list recursion, we again have a termination case and a general case. List recursion is usually appropriate for defining functions that require a list as one of its inputs and satisfies the following properties.

> **Termination Case** The value of the function is known immediately if the list is empty.
>
> **General Case** The value of the function can be computed by performing an operation that uses the value obtained by calling itself with the rest of the list as its input.

Here is an example of a function that uses list recursion to construct and return a list as its value:

Example 4.5.2 Define a function named **REVERSE**, which takes a list as its only input and returns a list whose elements have been reversed. For example,

```
REVERSE [1 2 3]      returns  [3 2 1]
REVERSE [A [X Y] C]  returns  [C [X Y] A]
REVERSE [ ]          returns  [ ]
```

The algorithm consists of the following cases:

Termination Case If the input list is empty, return the empty list.

General Case Construct a list by putting the last element onto the front of a list constructed by reversing the list containing all but the last element of the input list.

The definition of **REVERSE** is:

```
TO REVERSE :LIST
IF EMPTYP :LIST
   [OP [ ]]
   [OP FPUT LAST :LIST REVERSE BUTLAST :LIST]
END
```

Exercise 4.5.1. Define a function named **LENGTH** that has a list as its single input value. **LENGTH** returns a number that is the length of the list. For example, **LENGTH [A B C D]** returns **4**.

Exercise 4.5.2. Define a function named **MULTIPLY** that has a list of numbers as its only input. It returns the product of all the elements in the list.

Exercise 4.5.3. Use counting recursion to define a function named **ADDN** that has two inputs. **ADDN** returns a partial sum of the numbers in a list. The first input is a list of numbers and the second is a number indicating how many elements of the list to add. For example, **ADDN [2 4 7 8 1] 3** returns **13**.

Exercise 4.5.4. Describe the evaluation of **AVERAGE [4 5 9]** using substitution and simplification.

† **Exercise 4.5.5.** Define a function named **IS.MEMBERP** that requires two input values. The first input value is either a word or a list, and the second is a list. **IS.MEMBERP** returns **TRUE** if the first input value is an element of the second and **FALSE** otherwise. For example,

IS.MEMBERP "X [R T X D]	returns	TRUE
IS.MEMBERP "A [R T X D]	returns	FALSE
IS.MEMBERP "A [[A B] C D]	returns	FALSE
IS.MEMBERP [A B] [[A B] C D]	returns	TRUE
IS.MEMBERP "X [ ]	returns	FALSE

† **Exercise 4.5.6.** Define a function named **MEMBERN**, which requires two inputs as described for **IS.MEMBERP**. If the first input value is an element of the second input value, **MEMBERN** returns an integer representing its leftmost position in the list. If the first input is not a member of the list input, zero is returned. For example,

MEMBERN "X [R T X D] returns 3
MEMBERN [A B] [X Y Z [A B]] returns 4
MEMBERN "X [ ] returns 0

Exercise 4.5.7. Using substitution and simplification, describe the evaluation of each of the following expressions:

a. MEMBERN "X [R T X C]
b. REVERSE [A B C D]
c. REVERSE [A [B C D] E]
d. IS.MEMBERP "X [R T X D]
e. MEMBERN "X [[X Y] W Z]
f. IS.MEMBERP [A B] [X Y [R [A B]]]

4.6 Piecewise-Defined Functions

Many desired operations are defined piecewise. That is, the value of the function will be one thing if a certain condition is true, another thing if a second condition is true, and yet a third thing if a third condition is true.

The assumption here is that no more than one of the conditions will be true at the same time and that at least one will be true.

We can handle this situation by first making a test for one of the conditions. The true case is handled by computing and returning the appropriate value. The remaining conditions are handled by another IF.

Example 4.6.1. Define a Logo function that computes the value of the following function:

$$\text{sign}(x) = \begin{cases} 1 & \text{if } x > 0 \\ 0 & \text{if } x = 0 \\ -1 & \text{if } x < 0 \end{cases}$$

The Logo definition can be written as:

```
TO SIGN :X
IF :X > 0
  [OP 1]
  [IF :X = 0 [OP 0] [OP -1] ]
END
```

This technique for defining functions that have more than two pieces is referred to as **nesting if expressions**. This definition will work just fine, but it is harder to read. As Exercise 4.6.1 clearly illustrates, functions with even more pieces are even harder to read.

There are many operations that will have a number of cases to consider. What we need is a cleaner way of defining such functions. The type of function definitions we want to clean up are as follows:

1. The value of the function is defined in more than two pieces.

2. Each piece is defined by a test and a computation in which only one of the tests has a value of true.

The value to be returned by the function is the value of the computation associated with the first test that returns **TRUE**.

What we have in this situation is a collection of tests and a computation associated with each test. We evaluate the tests in the order presented. When we discover a test that is evaluated as true, we return the value computed by the computation associated with that test. The remaining tests, of course, are ignored.

A sequence of **IF**s, each with only two input expressions, does the job for us. For example, the **SIGN** function can be defined as:

```
TO SIGN :X
IF :X < 0 [OP -1]
IF :X = 0 [OP 0]
IF :X > 0 [OP 1]
END
```

Example 4.6.2 Define a function, named **STRIP**, which requires a word as its only input. If the first and last characters of the input are identical, **STRIP** returns a word formed by removing the first and last characters of the word. Otherwise, it returns a word that is the same as its input.

Algorithm This example is similar to a previous example, but the restriction on the length of the word has been lifted. The lifting of the length restriction means that we need to decide what to do if the input word is the empty word or a single-character word. In the first case, we should return the empty word. If the word is a single character, then the first character is certainly the same as the last character. Removing the first and last characters will result in the empty word again. We can specify the algorithm as follows:

1. If the input is empty (a word with zero characters), then return the empty word.

2. If the input is a single-character word, then return the empty word.

3. If the first and last characters of the input are the same, then return the word formed by removing the first and last characters of the input.

4. Otherwise, return the word itself.

Logo Definition

```
TO STRIP :OBJECT
IF :OBJECT = " [OP " ]
IF (LENGTH :OBJECT) = 1 [OP " ]
IF (FIRST :OBJECT) = LAST :OBJECT [OP BF BL :OBJECT]
OP :OBJECT
END
```

If the first three tests fail then we simply return the word itself.

Since the first two tests result in returning the same object (the empty word), we could combine the two tests into a single test and write the definition of **STRIP** as:

```
TO STRIP :OBJECT
IF OR :OBJECT = " (LENGTH :OBJECT) = 1 [OP " ]
IF (FIRST :OBJECT) = LAST :OBJECT [OP BF BL :OBJECT]
OP :OBJECT
END
```

Exercise 4.6.1. Define a Logo function that computes the value of the following function for any value of x using nested **IFS**.

$$f(x) = \begin{cases} 1 & \text{if } x < 0 \\ 2 & \text{if } 0 < x \leq 1 \\ 1 & \text{if } 1 < x < 2 \\ 0 & \text{if } 2 \leq x \leq 3 \\ 1 & \text{if } x > 3 \end{cases}$$

Exercise 4.6.2. Define a Logo function named **PALINDROMEP**, which requires a word as its single input. **PALINDROMEP** returns **TRUE** if its input is a palindrome and **FALSE** otherwise. A *palindrome* is a word that is the same if spelled backward. For example,

```
PALINDROMEP "LEVEL    returns        TRUE
PALINDROMEP "UNLEVEL  returns        FALSE
```

4.7 Stopping Recursive Turtle Procedures

We used recursion in Chapter 2 to define procedures that resulted in the turtle drawing a picture fragment forever. We call this technique **forever recursion** because the turtle will happily perform its antics forever. In this section, we introduce a procedure that allows us to halt the turtle at the appropriate time and also shows that counting and list recursion can be effectively used in defining recursive procedures for the turtle. We begin with an example.

Example 4.7.1 Define a procedure that causes the turtle to draw a spiral.

To cause the turtle to draw a spiral we start by asking it to move forward a certain distance and then turn a certain amount. We then ask it to repeat the same maneuver over and over with a shorter distance to move each time. That is, we can define this procedure as follows:

```
TO SPIRAL :DISTANCE :ANGLE
FD :DISTANCE RT :ANGLE
SPIRAL :DISTANCE - 5 :ANGLE
END
```

One problem with **SPIRAL** is that it never terminates. If we watch what happens when we execute

```
SPIRAL 100 90
```

we see that the turtle spirals inward and then starts retracing the spiral by moving backward over what it has drawn; it finally backs itself right off the edge of the screen.

The problem is that the distance moved forward eventually becomes negative. When this happens the **FD** statement becomes an instruction to move backward. Since the absolute value of the distance gets larger and larger, the turtle moves backward by ever-increasing distances.

If we want to draw a spiral and quit when the distance becomes non-positive, we need to check this condition before asking the turtle to do anything. We can define such a spiraling operation as follows:

```
TO SPIRAL.IN :DISTANCE :ANGLE
  IF :DISTANCE > 0 [ FD :DISTANCE RT :ANGLE
                     SPIRAL.IN :DISTANCE -5 :ANGLE]
END
```

Now, **SPIRAL.IN 100 90** will terminate when the value of :DISTANCE becomes non-positive. We can see what happens by tracing the evaluation of **SPIRAL.IN 20 90**:

```
SPIRAL.IN 20 90
=> IF 20 > 0 [FD 20 RT 90 SPIRAL.IN 20-5 90]
=> FD 20 RT 90 SPIRAL.IN 20-5 90
=> SPIRAL.IN 15 90
=> IF 15 > 0 [FD 15 RT 90 SPIRAL.IN 15-5 90]
=> FD 15 RT 90 SPIRAL.IN 15-5 90
=> SPIRAL.IN 10 90
   .
   .
   .
```

eventually arriving at:

```
=> SPIRAL.IN 5-5 90
=> IF 0 > 0 [FD 0 RT 90 SPIRAL.IN 0-5 90]
```

Since **0 > 0** is *false* and there is nothing else to do, **SPIRAL.IN** terminates after executing the following sequence of statements:

FD 20 RT 90 FD 15 RT 90 FD 10 RT 90 FD 5 RT 90

Although we do not need it to stop our spiraling procedure, there is another way to cause the inward-spiraling procedure to quit drawing. Since it will be required for more complex types of recursion we introduce it now.

> **STOP**
> **STOP** is a procedure that causes the execution of a user-defined procedure to terminate and return execution to the place from which it was called.

Example 4.7.2 Using **STOP**, we can define the **SPIRAL.IN** procedure as follows:

```
TO SPIRAL.IN :DISTANCE :ANGLE
IF :DISTANCE < 0 [STOP]
FD :DISTANCE
RT :ANGLE
SPIRAL.IN :DISTANCE - 5 :ANGLE
END
```

The type of recursion used depends on the size of one of its input values. This is true of counting recursion as well, but in counting recursion, the value of the counting variable decreases by a prescribed amount (one) and it is clear from the beginning how many times the operation will call itself for a specific value of the counting variable.

This definition of the **SPIRAL.IN** operation seems to do the same thing, but the value of :DISTANCE does not need to be an integer. Furthermore, the value of the variable :DISTANCE may change in totally unpredictable ways. Consider the following variation:

```
TO RANDOMSPIRAL :DISTANCE :ANGLE
IF :DISTANCE < 0 [STOP]
FD :DISTANCE
RT :ANGLE
RANDOMSPIRAL :DISTANCE - RANDOM 3 :ANGLE
END
```

In **RANDOMSPIRAL** we will not have a nice symmetric spiral. In fact, we might not even get a spiral. When it will stop is entirely unpredictable.

Exercise 4.7.1. Describe the sequence of statements sent to the turtle when the expression **SPIRAL 32 90** is executed.

Exercise 4.7.2. Define a Logo operation named **SPIRAL.OUT** that causes the turtle to draw an outward-growing spiral. That is, it starts with a short distance to move and increases for each recursive call.

Exercise 4.7.3. Define a Logo function that takes a point and a heading of 0, 90, 180, or 270 as input values and returns the distance from the point to the edge of the screen in the indicated direction.

† **Exercise 4.7.4.** Using the function you defined in Exercise 4.7.3, define an outward-spiraling operation that will cause the turtle to stop spiraling before it moves off the edge of the screen.

Exercise 4.7.5.* Define a Logo function that takes a point and a heading (any angle measure) as input values and returns the distance from the point to the edge of the screen in the indicated direction.

Exercise 4.7.6.* Using the function you defined in Exercise 4.7.5, define an outward-spiraling operation that will cause the turtle to stop spiraling just before it moves off the edge of the screen.

4.8 Mixing Turtles and Lists

Example 4.8.1 Define a procedure that requests the turtle to connect a collection of points together by drawing line segments joining consecutive points. The order in which the points are connected is determined by the order of the points.

> **Analysis** The problem is not completely defined, and we have to make some decisions on our own. The first question involves what to do if the collection is empty or has only a single element. If the collection contains no points, then we should do nothing at all. Or

---

* These exercises require the use of trigonometry.

should we? Before this question can be resolved, we need to decide what the state of the turtle should be when we are done drawing.

There are three reasonable choices here: leave the turtle at the last point with the idea that perhaps we will want to allow the turtle to continue wandering after it has connected all the points; return it to the home position; or return it to its position before we started connecting the points.

We choose the first alternative, to allow the turtle to continue its wanderings from the last point.

Algorithm Our decision to leave the turtle at the last point in the collection makes the decision of what to do if the collection contains zero or one point much easier. If there are zero points, we do nothing. If there is one point, we simply move the turtle to the desired point without drawing anything, since the single point is clearly the last point in the collection. Thus, our algorithm has the following form.

1. If there are no points, do nothing.

2. If there is at least one point, move the turtle to that point without drawing anything, and move the turtle through the remaining points with the pen down.

More Analysis We next need to decide how to represent a collection of points in Logo. The obvious choice is to represent the collection as a list, where each element is a point. Each point can be represented as a list of two numbers representing the point's x- and y- coordinates, respectively.

If we follow this algorithm in writing the Logo operation (let's call it **CONNECT.ALL**), we might arrive at the following definition using what we know about list recursion:

```
TO CONNECT.ALL :POINTS
IF EMPTYP :POINTS [STOP]
PENUP
```

```
SETPOS FIRST :POINTS
PENDOWN
CONNECT.ALL BUTFIRST :POINTS
END
```

If you now execute **CONNECT.ALL** you will immediately know that something is wrong. It should be clear from the behavior of the turtle what the problem is.

The last line is a recursive call to **CONNECT.ALL**, but **CONNECT.ALL** simply moves the turtle to the first point in the list without drawing and then does the same thing to the remaining elements of the list. That is not what we want to do at all. Lines 2, 3, and 4 initialize the turtle in preparation for connecting the remaining points. We must think of **CONNECT.ALL** as doing an initialization followed by a repetitive process. That is, the last line of the body should be a call to a different function. Our redesigned algorithm is:

1. If there are no points in the list, then do nothing.

2. Initialize the turtle by moving it to the first point in the list with the pen up.

3. Execute an operation that moves the turtle through the points contained in the rest of the list with the pen down.

The new Logo definition (let's call it **CONNECT.POINTS**) becomes

```
TO CONNECT.POINTS :POINTS
IF EMPTYP :POINTS [STOP]
INITIALIZE.TURTLE FIRST :POINTS
CONNECT.REST BUTFIRST :POINTS
END
```

where the initialization procedure is given by:

```
TO INITIALIZE.TURTLE :POINT
PENUP
SETPOS :POINT
PENDOWN
END
```

Mixing Turtles and Lists **131**

We view **CONNECT.REST** as a procedure that draws lines from wherever the turtle is positioned to the first point in the list, then draws a line to the next point in the list, and continues until all points have been connected. That is, the algorithm for **CONNECT.REST** is as follows:

1. If the collection of points is empty, stop.

2. Otherwise, draw a line to the first point in the list and execute **CONNECT.REST** to connect the remaining points in the list.

Our algorithm clearly meets the conditions for list recursion. The definition of **CONNECT.REST** is

```
TO CONNECT.REST :POINTS
IF EMPTYP :POINTS [STOP]
SETPOS FIRST :POINTS
CONNECT.REST BUTFIRST :POINTS
END
```

We have also created a way to continue the turtle's wanderings from the point it left off. Namely, simply call the operation **CONNECT.REST** with an additional list of points as its input value. We do not need to remember the last point in the collection given to **CONNECT.REST**.

Exercise 4.8.1. Define a procedure named **CONNECT.TO.POINT** that requires two input values. The first input is a list representing a point, which we call the central point. The second is a list of points. **CONNECT.TO.POINT** connects the central point to each point in the list of points.

† **Exercise 4.8.2.** Define a procedure named **CONNECT.AND.POSITION** that takes a list of points and a terminating point as its input values. **CONNECT.AND.POSITION** connects the list of points together (not including the turtle's initial position or the terminating point) and places the turtle at the terminating point.

Exercise 4.8.3. Define a procedure named **CONNECT.AND.RETURN** that takes a list of points as its only input value. The operation connects the points in the order they appear in the list and returns the turtle to its original position without connecting its starting position to any point in the list.

4.9 Graphing Functions

The graph of a function may be approximated on the graphics screen by generating a list of points and then using **CONNECT.POINTS** to connect them. For example, if we want to sketch the graph of $y = x^2 - 20x + 75$, where x ranges from 0 to 50, we could generate the list of points

   [[0 75] [5 0] [10 -25] ... [50 1575]]

and then connect them. Assuming that we have a function named **GENERATE.POINTS** that returns a list of points on the graph of the above equation, we could define a graphing procedure as follows:

    TO GRAPH
    WINDOW
    CONNECT.POINTS   GENERATE.POINTS
    END

We set the boundary type of the graphics screen to window mode because there may be points on the graph that will not be on the display. If it were in **FENCE** mode, an error would be generated if the turtle was sent off the screen. If it were in **WRAP** mode, we would get unpredictable and undesireable line segments on the screen.

The function **GENERATE.POINTS** currently has no inputs. It returns a list of points on the graph of the function we want to graph. This list of points is then used as the input to **CONNECT.POINTS**.

GENERATE.POINTS has to know the interval containing all the x-values in which we are interested, the increment between two successive values of x, and the function we are graphing. Since we want to use our graphing operation for other equations besides $y = x^2 - 20x + 75$ as well as use it for other intervals for x, we do not define **GENERATE.POINTS** with all this information built in.

To begin, we define a function named **Y**, which returns the value of y that satisfies the equation for a particular value of x. Thus, the definition of **Y** for the above equation is

```
TO Y :X
OP :X*:X - 20*:X + 75
END
```

Whenever we want to compute the value of y that satisfies the equation for a particular value of x, we just call **Y** with the value of x as its input. For example, to compute the y-value of the point whose x-value is 10, we type

```
Y 10
```

Whenever we want to graph a new function, we just redefine **Y**.

To generate a point on the graph of the equation we construct a list containing a value for x and the corresponding value of y that satisfies the equation. For example, to construct the point on the graph whose x-coordinate is 10, we use the following expression:

```
LIST 10  Y 10
```

(Note that [10 Y 10] is considered a constant by Logo and is not the same as the value constructed by **LIST 10 Y 10**.)

To generate a list of points on the graph, we need to build a list containing points whose x-coordinates range from the lowest desired value of x to the highest desired value of x. **GENERATE.POINTS** needs to know what these values are as well as the distance between two successive values of x. There are two ways to do this:

1. Have the lowest and highest values of x and the increment between successive x's be inputs to **GENERATE.POINTS**.

2. Define **GENERATE.POINTS** with no inputs and write the definition so that the values are known by the definition.

We choose the first alternative, since the second choice will force us to change the definition of **GENERATE.POINTS** every time we want to change the interval. This decision means we have to redefine the graphing function as follows:

```
TO GRAPH :LEFTEND :RIGHTEND :INCREMENT
WINDOW
CONNECT.POINTS  GENERATE.POINTS :LEFTEND
                                :RIGHTEND
                                :INCREMENT
END
```

The user will be expected to provide these values when a particular graph is desired.

We still need to define **GENERATE.POINTS**. The inputs will be the three values just described, so we will give them identical names. Thus, the first line of the definition looks like this:

TO GENERATE.POINTS :LEFTEND :RIGHTEND :INCREMENT

To construct the first point on the graph, we construct the point on the graph at the left end of the interval for x. That is, we compute

    MK.POINT :LEFTEND   (1)

where **MK.POINT** is defined by

```
TO MK.POINT :X
OP LIST :X Y :X
END
```

We want to add this point to a list of points obtained by letting x vary from **:LEFTEND + :INCREMENT** to **:RIGHTEND** using **:INCREMENT** as the distance between succesive x-values. **GENERATE.POINTS** will handle this job. That is, the list of remaining points is generated by calling **GENERATE.POINTS** recursively:

    GENERATE.POINTS :LEFTEND + :INCREMENT   (2)
                             :RIGHTEND
                             :INCREMENT

The entire list of points we want is obtained by putting the point constructed in (1) onto the front of the list of points constructed by (2):

```
FPUT MK.POINT :LEFTEND
    GENERATE.POINTS :LEFTEND + :INCREMENT
                    :RIGHTEND
                    :INCREMENT
```

The one remaining question is to decide when to terminate the recursion. Since the value of :LEFTEND gets larger with each recursive call, we should terminate the recursion when :LEFTEND becomes greater than :RIGHTEND. Now we need to decide what to return when :LEFTEND is greater than :RIGHTEND. Because GENERATE.POINTS is expected to return a list, the only reasonable thing to return is the empty list. Thus, we include the following terminating case:

```
IF :LEFTEND > :RIGHTEND [OP [ ]]
```

Putting the pieces together, we arrive at the following definition of GENERATE.POINTS:

```
TO GENERATE.POINTS :LEFTEND :RIGHTEND :INCREMENT
IF :LEFTEND > :RIGHTEND
   [OP []]
   [OP FPUT MK.POINT :LEFTEND
           GENERATE.POINTS :LEFTEND + :INCREMENT
                           :RIGHTEND
                           :INCREMENT]
END
```

As mentioned earlier, parts of the graph may not lie on the screen. There are two ways we can think about viewing the missing parts:

1. We can try to move the graph up, down, to the left, or to the right by a simple translation. This means that part of the graph now visible will disappear while other parts become visible.

2. If we wish the entire graph to appear on the screen, we can scale the points generated by GENERATE.POINTS so that the entire graph appears. That is, we transform all the "real" points on the graph into "displayable" points.

For example, if we want to move the graph down a distance of 100 we simply subtract 100 from each of the *y*-coordinates in the list of points generated by **GENERATE.POINTS**.

Assuming that we have operations named **MOVEDOWN**, **MOVEUP**, **MOVELEFT**, and **MOVERIGHT** that translate a list of points up, down, left, or right by a prescribed amount and functions named **SCALE.X** and **SCALE.Y** that scale the *x*- or *y*-coordinates of a list of points, let's define an interactive program that draws the graph of a function and then allows the user to translate or scale the graph at his or her command. The top-level graphing operation is named **EXPLORE.GRAPH**. It takes the same inputs as **GRAPH** does, puts the graphics screen in window mode, generates the points on the graph and then calls **DRAW.GRAPH**.

```
TO EXPLORE.GRAPH :LEFTEND :RIGHTEND :INCREMENT
WINDOW
DRAW.GRAPH GENERATE.POINTS :LEFTEND
                          :RIGHTEND
                          :INCREMENT
END
```

DRAW.GRAPH clears the screen, connects the points, and then calls itself with a new set of points resulting from translating the current set of points. The translation is done by calling a function named **MODIFY.GRAPH**. The inputs to **MODIFY.GRAPH** are the current list of points and a character indicating which direction the graph should be translated.

```
TO DRAW.GRAPH :POINTS
CLEARSCREEN
CONNECT.POINTS :POINTS
DRAW.GRAPH MODIFY.GRAPH :POINTS READCHAR
END
```

The character is obtained from the user by the **READCHAR** operation described in Chapter 2:

> **READCHAR** is a function with no input values. If a character has been typed on the keyboard, that character is returned. If no character has been typed, **READCHAR** will wait until the user types one.

If the character typed is a *u* the graph is moved up by 40 units. If the character is a *d* the graph is moved down by that amount. *l* and *r* control movement to the left and right, respectively. If any other character is typed, **MODIFY.GRAPH** will read another character and call itself with the unchanged list of points.

```
TO  MODIFY.GRAPH :POINTS :CHR
IF :CHR = "D [OP MOVEDOWN :POINTS 40]
IF :CHR = "U [OP MOVEUP :POINTS 40]
IF :CHR = "L [OP MOVELEFT :POINTS 40]
IF :CHR = "R [OP MOVERIGHT :POINTS 40]
            [OP MODIFY.GRAPH :POINTS  READCHAR]
END
```

† **Exercise 4.9.1.** Define a function named **MOVEDOWN** that takes a list of points and a number as its input values. **MOVEDOWN** returns a list of points in which the y-coordinate of each point in the list has been decreased by the amount specified by the number.

Exercise 4.9.2. Define a function named **MOVEUP** that takes a list of points and a number as its input values. It returns a list of points formed by increasing the y-coordinate of each point in the list by a specified amount. (*Hint*: Use the operation defined in Exercise 4.9.1.)

Exercise 4.9.3. Define a function named **MOVELEFT** that takes a list of points and a number as its input values. It returns a list of points formed by decreasing the x-coordinate of each point in the list by a specified amount.

Exercise 4.9.4. Define a function named **MOVERIGHT** that takes a list of points and a number as its input values. It returns a list of points formed by increasing the x-coordinate of each point in the list by a specified amount. (*Hint*: Use the operation defined in Exercise 4.9.3.)

† **Exercise 4.9.5.** Define a function named **SCALE.X** that requires a list of points as its only input. It returns a list of points representing the transformation of the x-coordinates of its input list to the interval from -100 to 100.

Exercise 4.9.6. Define a function named **SCALE.Y** that requires a list of points as its only input. It returns a list of points representing the transformation of the *y*-coordinates of its input list to an interval from -70 to 100.

† **Exercise 4.9.7.** Using Exercises **4.9.5** and **4.9.6**, define a function named **SCALED.GRAPH** by modifying **GRAPH** so that it displays a scaled graph of a function. That is, it scales the graph so that it will fit nicely on the screen.

† **Exercise 4.9.8.** Modify **EXPLORE.GRAPH** so that it will terminate when the user types a *q*.

4.10 More on Graphing Functions

We want to use the graphing operations created in Section 4.9 to graph many different functions. To do so, we have to change the definition of *y* each time. To make graphing different functions more natural, we rewrite the graphing operation to admit the function we are graphing as an input to **GRAPH** along with **:LEFTEND**, **:RIGHTEND**, and **:INCREMENT**. This eliminates the need to redefine **Y**.

Our assumption is that we want to graph an equation of the form

    Y = *an expression in x*

Our earlier example would be written

    Y = X∗X - 20 ∗ X + 75

Since *y* = will always be there, the information that the program needs is described by the expression to the right of the equals sign. If we provide a fourth input variable, to **GRAPH**,

```
TO GRAPH :LEFTEND :RIGHTEND :INCREMENT :FUNCTION
WINDOW
CONNECT.POINTS GENERATE.POINTS :LEFTEND
                                :RIGHTEND
                                :INCREMENT
                                :FUNCTION
END
```

(assuming that **GENERATE.POINTS** has been appropriately modified to handle the function input) and type in the expression

GRAPH 0 50 5 X * X - 20 * X + 75

Logo would try to evaluate **X*X - 20*X + 75** before ever calling **GRAPH** and we would wind up with a nasty error message. The problem is that **X*X - 20*X + 75** is viewed by Logo as an expression that should be evaluated. Preventing the evaluation of **X*X - 20*X + 75** can be solved by packaging the function into a list and calling **GRAPH** as follows:

GRAPH 0 50 5 [X * X - 20 * X + 75]

Now the right-hand side of the equation is considered to be a list of words, and no attempt to evaluate its contents is made. The list [X * X - 20 * X + 75] becomes the value of the **:FUNCTION** input variable.

We now turn our attention to the problem of how to modify **GENERATE.POINTS** so that it will use the list representing the function to compute the value of y that is paired with a particular value of x. Recalling the previous definition of **GENERATE.POINTS**,

```
TO GENERATE.POINTS :LEFTEND :RIGHTEND :INCREMENT
IF :LEFTEND > :RIGHTEND
   [OP []]
   [OP FPUT MK.POINT :LEFTEND
            GENERATE.POINTS :LEFTEND + :INCREMENT
                            :RIGHTEND
                            :INCREMENT]
END
```

we see that **MK.POINT** needs the function in order to construct the point on the graph. We must add the function input in the recursive call to **GENERATE.POINTS**. The new definition of **GENERATE.POINTS** becomes:

```
TO GENERATE.POINTS :LEFTEND :RIGHTEND :INCREMENT :FUNCTION
IF :LEFTEND > :RIGHTEND
   [OP []]
   [OP FPUT MK.POINT :LEFTEND :FUNCTION
           GENERATE.POINTS :LEFTEND + :INCREMENT
                           :RIGHTEND
                           :INCREMENT
                           :FUNCTION]
END
```

We must now modify the definition of **MK.POINT**. Isolating the expression in the definition of **MK.POINT** that computes the value of the function at **:X**,

Y :X

we must replace it by one that will compute the y-value using the **:FUNCTION** input as well. We define a Logo operation named **EVALFUN** to perform this computation. That is, we replace the expression **Y :X** by **EVALFUN :FUNCTION :X**. The new definition of **MK.POINT** is

```
TO MK.POINT :X :FUNCTION
OP LIST :X :EVALFUN :FUNCTION :X
END
```

EVALFUN takes a list representing the function and a value for x as inputs. What we need to do is clear. We need to substitute the value of x for every occurrence of the word x in the list and then evaluate the expression in the list. That is, the definition of **EVALFUN** is

```
TO EVALFUN :FUNCTION :X
OP RUN (REPLACE.ALL "X :FUNCTION :X)
END
```

The definition of **REPLACE.ALL** is left as an exercise, and the **RUN** primitive was described in Chapter 2.

> **RUN** *list*
> **RUN** is a Logo operation that takes a list as its only input value. The expressions represented by the elements of the list are executed.

† **Exercise 4.10.1.** Define a function named **REPLACE.ALL** that takes a word, **W**, a list, **L**, and a word, **E**, as inputs. It returns a list identical to **L** except that all elements of the list that are identical to **W** are replaced by **E**. For example,

REPLACE.ALL "A [A B A C A] "Q
returns the list [Q B Q C Q]

REPLACE.ALL "X [X * X + 3 * X] 5
returns the list [5 * 5 + 3 * 5]

Exercise 4.10.2. Modify **EXPLORE.GRAPH** so that it requires the function to graph as an input.

Exercise 4.10.3. Modify **SCALED.GRAPH** (see Exercise 4.9.7) so that it requires the function to graph as an input.

Chapter 5
Level Diagrams: Another Model For Describing Computation

5.1 Introduction

This chapter introduces a graphical model for describing how a computation proceeds as well as for exploring recursion in its most general form. The graphical model consists of building a collection of boxes, called frames. Each frame represents a call to a user-defined function or procedure. The collection of frames is called a **level diagram** and represents a complete history of who called whom at what time with what input values. That is, it represents the **calling chain** of user-defined functions and procedures.

After introducing level diagrams, a number of new operations dealing with lists are defined. They are used to help in the solution of problems described in later chapters as well as providing additional examples to model with the use of level diagrams.

The remaining material is aimed at understanding recursion in its most general form.

5.2 Level Diagrams

Models of computation provide a way of examining a computation without looking at what the machine does when it executes a statement. That is, they are mind activities rather than machine activities. A model is not an imitation of a machine but a description of how a machine could perform a computation. Of course, a model must provide an accurate description, in that it should provide the same result as the machine in all computations.

Level Diagrams 143

Having more than one model is useful because one model may provide more insight into what is going on than another in a particular situation. Successful debugging strategies use all the insight that can be mustered. The substitution and simplification model is entirely adequate for describing any computation that you have seen so far or that you will see in this chapter, but the level-diagram model allows us to view the most general forms of recursion in a new light.

When we define a new procedure, such as **RECTANGLE**, Logo tucks it away in the workspace. If we now type in

RECTANGLE 25 50

a copy of the definition of **RECTANGLE** is obtained from the workspace. But instead of substituting the input values for **:WIDTH** and **:LENGTH** in the body of this copy as described in the substitution and simplification model, the values of the input expressions (**25** and **50**) are paired with the names of the input variables (**WIDTH** and **LENGTH**). The pairing of a value to the name of a variable is called a **variable binding**.

The statements in the body of **RECTANGLE** are then executed. Any time an input variable is encountered during the execution of a statement, the Logo evaluator looks up the value that is currently bound to the named variable. That is, the substitution of a variable by its value is performed only when it is needed.

Thus, we can imagine a more streamlined execution of the bodies of user-defined functions and procedures. Instead of examining the body twice (once for substituting values for the input variables and once when the statements in the body are executed), the evaluator simply starts executing the statements. When an input variable is encountered, it looks up the value currently bound to it.

We can describe the streamlined execution of **RECTANGLE 25 50** by the connected boxes in Figure 5.1.

144 Level Diagrams: Another Model For Describing Computation

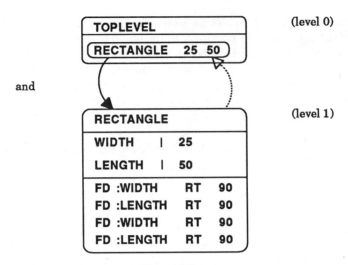

Figure 5.1

Each box is called a **frame.** The first frame has the title **TOPLEVEL** and contains the statement **RECTANGLE 25 50**. This indicates what is requested by the user and is called the **toplevel frame**. The title of the second frame is the name of the user-defined procedure called by the statement in the toplevel frame. It contains the values that have been bound to the input variable names and the statements that make up the body of **RECTANGLE**. These bindings remain unchanged while executing the statements contained in the lower portion of the frame (i.e., the body of the **RECTANGLE** procedure).

The relationship between the frames is indicated by drawing a loop around the entire procedure call in the toplevel frame with an arrow pointing to the new frame. This indicates that in order to complete the execution of **RECTANGLE 25 50**, the execution of the statements in the other frame must be completed. The arrow pointing back to the toplevel frame indicates that control is passed back up to the toplevel frame when the call to **RECTANGLE** has finished.

The toplevel frame indicates what the user types in when prompted by the Logo system. It is also called the level-zero computation. Any frame resulting from a level-zero computation is called a level one frame, any frame resulting from a level-one computation is called a level-two frame, and so on. The creation of another frame at a lower level is analogous to replacing a call to a user-defined function or procedure by a copy of its body with the appropriate substitution for input varibles in the substitution and simplification model.

When the execution of the body in the level-one frame has finished, control is returned to the toplevel frame, discarding the level one frame (including its variable bindings). Since this completes the execution of the toplevel statement, another ? prompt is displayed, and Logo waits for a new request from the user.

Once we define a new function or procedure, its definition is seldom changed, whereas the values associated with variables change quite frequently. In fact, as we see shortly, a variable name may have several different values bound to it at any given time during the execution of a toplevel statement. The following examples show how this happens and the convention Logo uses to decide which variable binding is used during execution of a statement.

We first describe how level diagrams are used in the execution of a statement that calls a user-defined function. As an example, the execution of **SHOW SQ 4** is described by the level diagram in Figure 5.2.

The toplevel statement causes the creation of the level-one frame named **SQ**, because **SQ** is a user-defined function. The execution of the toplevel statement must wait until the level-one computation completes. The level-one frame says to compute the product of **:X** with itself with **:X** bound to **4** and to **OUTPUT** the result. **OUTPUT**ting a result means replacing the looped expression by the value computed in the pointed to frame. That is, **SQ 4** is replaced by (simplified to) the value produced by the level-one computation. The result is then displayed by **SHOW**.

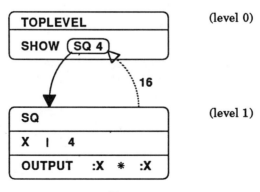

Figure 5.2

146 *Level Diagrams: Another Model For Describing Computation*

The arrow pointing back to the looped expression in the toplevel frame with the number **16** next to it indicates the value that is to replace the looped expression.

A more complex computation is the execution of **SHOW SUMSQ 3 5** at the toplevel, where **SUMSQ** is defined as before. The level diagram describing this computation is given in Figure 5.3.

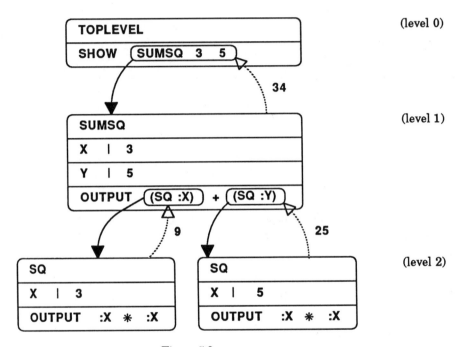

Figure 5.3

The inputs to **SUMSQ** are **3** and **5**. The level-one frame is created, and execution of the statement in the body of **SUMSQ** with **3** bound to **X** and **5** bound to **Y** is begun. During the execution, the evaluator sees that **SQ :X** must be simplified. The input expression to **SQ** is simplified obtaining the value **3** (the value currently bound to **X**). A level-two computation (the left level-two frame) is created that represents the computation of **SQ** with **3** bound to **X**. This computation will return **9** as the value of the first input to **+** in **SUMSQ**. Back at level one the execution of **SQ :Y** is performed. The value of **:Y** is **5** and another level-two computation is created (the right level-two frame) to perform the execution of **SQ** with **5** bound to **X**.

The left level-two frame has been discarded, because that computation was completed long ago. Thus we are left with two bindings for the variable named **X**. When the evaluator is busy executing the statements in a particular frame, the variable binding used is the one in that frame. That is, it uses the binding at level two.

The convention used to find the value of a variable is summarized next. It is called a **scoping rule**. The particular rule used by Logo is called **dynamic scoping**. A complete discussion of the implications of this rule is given in Chapter 6.

Dynamic Scoping Rule

When evaluating a statement belonging to a particular frame, the binding used in determining the value of a variable is the one in the same frame.

If no such binding exists in that frame, look for the binding in the frame at the previous level. Continue this backward search until a binding is found.

Exercise 5.2.1. Draw the level diagram describing the execution of the following toplevel statements:

   a. **RECTANGLE 50 80**
   b. **DEFINEPOLY 5** (See definition in Section 3.9.)
   c. **POLY5 20**

Exercise 5.2.2. Draw the level diagram describing the execution of **HOW.FAR**, assuming that the Turtle is currently located at the point (3, 5). The definition of **HOW.FAR** is given by:

   TO HOW.FAR
   OUTPUT SQRT SUMSQ XCOR YCOR
   END

5.3 Modeling Recursion with Level Diagrams

This section presents three examples of using level diagrams in the description of recursive procedures and functions.

Example 5.3.1. Describe the execution of **DOSQUARE 50** using level diagrams.

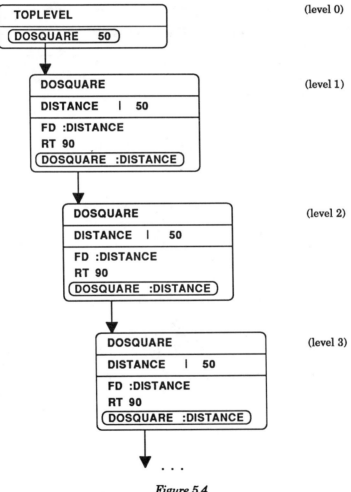

Figure 5.4

At the toplevel, the **DOSQUARE** procedure is called with **50** as its input (Figure 5.4). This creates the level-one computation with **50** bound to the input variable **:DISTANCE**. At level one the turtle is moved forward 50 and turned right 90° before calling **DOSQUARE** with an input of **50**. The call to **DOSQUARE** in the level-one frame creates the level-two computation. This sequence repeats itself at level two, creating the level-three computation. This continues "forever." None of the lower-level computations is ever completed; we just keep adding new ones.

Example 5.3.2 Describe the execution of **SHOW POWER 4 3** using level diagrams.

See Figure 5.5. The top-level frame indicates that we are evaluating **SHOW POWER 4 3**. The input values to power (**4** and **3**) are bound to **X** and **N** in the level-one frame and the **IF** statement in the body of **POWER** is executed.

In order to complete the level-one computation, we must compute the value of **POWER :X :N-1**. We first evaluate the input expressions to **POWER** at level one (we find the values of **:X** and **:N-1** to be **4** and **2**, respectively). Again, since **POWER** is a user-defined function, we create a new computation (indicated by the level-two frame) by binding the values **4** and **2** with the names of the input variables of **POWER**. Again, the **IF** statement in the body of **POWER** is evaluated.

At this point we have three levels of computations, none of which have completed. The process continues until we have created and begun to execute the **IF** statement at level four. The level-four computation completes since **:N** has value zero. It outputs the value **1** to the level-three computation by replacing the looped level-three expression with **1**.

150 Level Diagrams: Another Model For Describing Computation

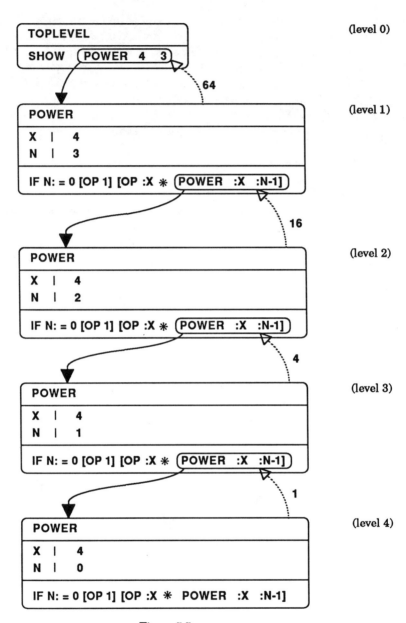

Figure 5.5

The level-three computation can now be completed by multiplying its binding of :X (4) with the result returned from level four (1), and replacing the looped level-two expression by 4.

The level-two computation may now complete its work by replacing the indicated loop in level one by 16.

The level-one computation completes by replacing the circled expression in the toplevel with 64. At the toplevel, the number 64 is displayed.

Note that the level diagrams parallel what happens in the substitution and simplification model. New computations are created and the number of levels grows until :N becomes zero. This corresponds to the expansion of the computation in the substitution and simplification scheme. When :N becomes zero, we return to the computation of the previous level, which in turn completes its work and returns to the level above that. These steps correspond to the shrinking of the expressions at each step in the substitution and simplification model of computation.

Example 5.3.3 Describe the execution of **SHOW ADD [3 5 7]** using level diagrams. See Figure 5.6.

152 Level Diagrams: Another Model For Describing Computation

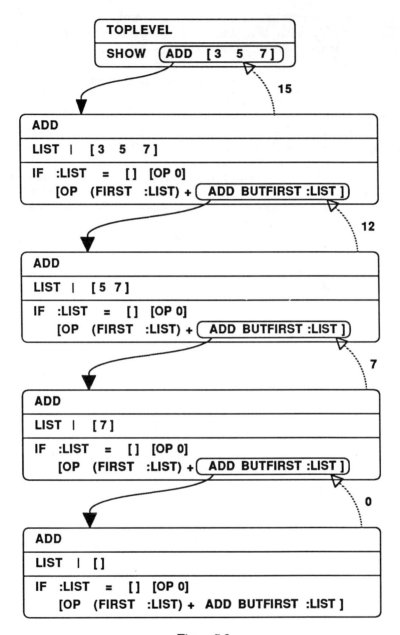

Figure 5.6

Exercise 5.3.1. Using level diagrams, describe the execution of **SPI 4 90 20** defined in Exercise 3.7.6.

Exercise 5.3.2. Using level diagrams, describe the execution of **SHOW FACT 3**, where **FACT** is defined by

```
TO FACT :N
IF :N = 0 [OP 1] [OP :N * FACT :N-1]
END
```

Exercise 5.3.3. Draw the level diagram describing the execution of **SHOW NTH [A B C D] 3** using the definition of **NTH** in Example 4.4.2.

Exercise 5.3.4. Draw the level diagram describing the execution of **CONNECT.POINTS [[10 10] [20 20] [30 30]]** using the definition of **CONNECT.POINTS** in Section 4.8.

5.4 Some List Utility Functions

Logo provides several primitive functions to be used with lists, which can be categorized as selectors, constructors, or recognizers.

Selectors	*Constructors*	*Recognizers*
FIRST	**FPUT**	**LISTP**
BUTFIRST	**LIST**	**WORDP**
ITEM	**SENTENCE**	**EMPTYP**

We want to add our own functions to this collection of primitives. For example, in addition to **BUTFIRST** and **BUTLAST**, it would be useful to have a selector named **BUTNTH** that takes a positive integer, n, and a list as inputs and returns a list with the nth element of its list input removed. For example,

BUTNTH 3 [A B C D E]

returns the list

[A B D E].

Example 5.4.1 Define a selector named **BUTNTH**, which requires a positive integer, n, and a list as its input values and which returns a list with the nth element removed.

The definition begins with

TO BUTNTH :N :LIST

Then, we reason as follows:

1. If the list is empty, then return the empty list. (The empty list with its nth element removed is still the empty list.) We now have

TO BUTNTH :N :LIST
IF EMPTYP :LIST [OP [ ]]
...
END

2. If the list is not empty and the value of the integer is 1, then return **BUTFIRST** of the list. (That is, throw out the first element.) Adding this line, we have

TO BUTNTH :N :LIST
IF EMPTYP :LIST [OP [ ]]
IF :N = 1 [OP BUTFIRST :LIST]
...
END

We cannot continue this process indefinitely, because we will never run out of positive integers. We must decide what to do when **:N** has a value other than 1.

3. If the list is not empty and the value of the integer is not equal to 1, return a list constructed as follows:

    a. Compute **BUTNTH :N-1 BUTFIRST :LIST**.
    b. **FPUT** the first element of **:LIST** onto the
    result obtained in (a).

We can now complete the definition of **BUTNTH:**

    to BUTNTH :N :LIST
    IF EMPTYP :LIST [OP [ ]]
    IF :N = 1 [OP BUTFIRST :LIST]
       [OP FPUT FIRST :LIST BUTNTH :N-1 BUTFIRST :LIST]
    END

Example 5.4.2 Define a constructor named **REPLACE.NTH** that requires a number, a list, and an object (a word or a list) as input values. It returns a list in which the *n*th element of the list input has been replaced by the object input. For example,

    REPLACE.NTH 3 [A B C B] "Z returns **[A B Z B]**
    REPLACE.NTH 7 [A B C B] "Z returns **[A B C B]**
    REPLACE.NTH 3 [ ] "Z returns **[ ]**

The skeleton of the definition has the form

    TO REPLACE.NTH :N :LIST :OBJ
       ...
    END

The body is very similar to **BUTNTH:**

1. If the list is empty, then return the empty list:

    TO REPLACE.NTH :N :LIST :OBJ
    IF EMPTYP :LIST [OP [ ]]
       ...
    END

2. If **:N = 1,** put the new object onto the front of the rest of the list:

```
TO REPLACE.NTH :N :LIST :OBJ
IF EMPTYP :LIST [OP [ ]]
IF :N = 1 [OP FPUT :OBJ BUTFIRST :LIST]
...
END
```

3. If :N is not equal to 1, then do the following:

    a. Compute **REPLACE.NTH :N-1 BF :LIST :OBJ**.
    b. **FPUT FIRST :LIST** onto result in (a).

```
TO REPLACE.NTH :N :LIST :OBJ
IF EMPTYP :LIST [OP [ ]]
IF :N = 1 [OP FPUT :OBJ BUTFIRST :LIST]
   [OP FPUT FIRST :LIST
              REPLACE.NTH :N-1 BF :LIST :OBJ]
END
```

Exercise 5.4.1. Describe the execution of each of the following using simplification and substitution:

   † a. **SHOW BUTNTH 3 [A B C D E]**
      b. **SHOW BUTNTH 2 [A [B C D] E]**
      c. **SHOW BUTNTH -3 [A B C D E]**
      d. **SHOW BUTNTH 7 [A B C]**

Exercise 5.4.2. Describe the execution of each of the following using level diagrams:

   † a. **SHOW BUTNTH 3 [A B C D E]**
      b. **SHOW BUTNTH 2 [A [B C D] E]**
      c. **SHOW BUTNTH -3 [A B C D E]**
      d. **SHOW BUTNTH 7 [A B C]**

Exercise 5.4.3. Define a binary function named **ALL.BUT.FIRST**. The first input is a list of words and the second is a word. It returns a copy of the list input with the first occurrence of the word input omitted. For example,

    ALL.BUT.FIRST [A B C B D] "B WILL return **[A C B D]**
    ALL.BUT.FIRST [A B C D B] "E WILL return **[A B C D B]**

Exercise 5.4.4. Using your answer to Exercise 5.4.3, what values (or errors) are returned by the following expressions?

   a. **ALL.BUT.FIRST [A [B C] B D] "B**
   b. **ALL.BUT.FIRST [A [B C] B D] [B D]**
   c. **ALL.BUT.FIRST [A [B [C B]] D] "B**

Exercise 5.4.5. Define a function named **ALL.BUT** that requires two input values. The first input is assumed to be a list and the second input is either a word or a list. It returns a list identical to the list input with all list elements that are the same as the second input omitted. For example,

   ALL.BUT [A B C B D] "B will return **[A C D]**
   ALL.BUT [A [B C] D B] "B will return **[A [B C] D]**

† **Exercise 5.4.6.** Describe the execution of **ALL.BUT [A [A B C] X A B] "A** using simplification and substitution.

Exercise 5.4.7. Describe the execution of **ALL.BUT [A [A B C] X A B] "A** using level diagrams.

† **Exercise 5.4.8.** Define a function named **REPLACE.FIRST** that requires a list and two objects (words or lists) as input values and returns a list in which the first occurrence of an element of the input list that matches the first input object is replaced by the second object input. For example,

   REPLACE.FIRST [A B C A] "A 3 returns **[3 B C A]**

Exercise 5.4.9. Define a function named **REPLACE.ALL** that requires a list and two objects (words or lists) as input values and returns a list in which every occurrence of an element of the input list that matches the first object is replaced by the second object. For example,

   REPLACE.ALL [A B C A] "A 3 returns **[3 B C 3]**
   REPLACE.ALL [A B [C A]] "A 3 returns **[3 B [C A]]**
   REPLACE.ALL [A B C A] "Z 3 returns **[A B C A]**

5.5 Tree Recursion

In this section we look at two examples of functions and one example of a procedure whose bodies contain more than one recursive call. That is, we look at recursion in its most general form.

Example 5.5.1 An interesting sequence of numbers is the Fibonacci sequence. The first two elements of the sequence are 1 and the remaining elements are formed by adding together the previous two. The first few numbers in the sequence are:

n	0	1	2	3	4	5	6	7	8	9	...
fib (n)	1	1	2	3	5	8	13	21	34	55	...

Define a function named *fib* that will return the nth number in this sequence, where the zeroth and first numbers are both 1. That is, *fib* takes a nonnegative integer, n, as an input and returns the nth number in the sequence.

The definition of the *fib* function can be made more compact and precise by writing it as follows:

$$\text{fib}(n) = \begin{cases} 1 & \text{if } n = 0 \\ 1 & \text{if } n = 1 \\ \text{fib}(n-1) + \text{fib}(n-2) & \text{if } n > 1 \end{cases}$$

We can transform this definition into Logo as follows:

```
TO FIB :N
IF :N = 0 [OP 1]
IF :N = 1 [OP 1]
   [OP (FIB :N-1) + (FIB :N-2)]
END
```

The definition of **FIB** is recursive, but it differs from previous examples in that it has two calls to itself. The form (**FIB :N-1**) + (**FIB :N-2**) is evaluated by first computing the value of **FIB :N-1**, then computing the value of **FIB :N-2**, and finally adding the results together.

It is important to remember that **FIB :N-1** and **FIB :N-2** are two entirely separate computations. We can picture what is happening by looking at the level diagrams for the execution of **SHOW FIB 4** in Figure 5.7.

The computation of **SHOW FIB 4** is treelike in behavior, but there is never any more than one active computation at each level. For example, if we have just begun the evaluation of the body of **FIB** with **:N** bound to 1 at level four, the only computations active are the leftmost frames on all levels.

If we have just begun the level-two computation with **N** bound to **2**, the active computations are the rightmost frames at levels zero, one, and two. Note that **FIB 2** has already been computed at level three, but this result is no longer available when we need to compute **FIB 2** at level two. This is an inefficiency in the algorithm we are using to compute the function.

160 *Level Diagrams: Another Model For Describing Computation*

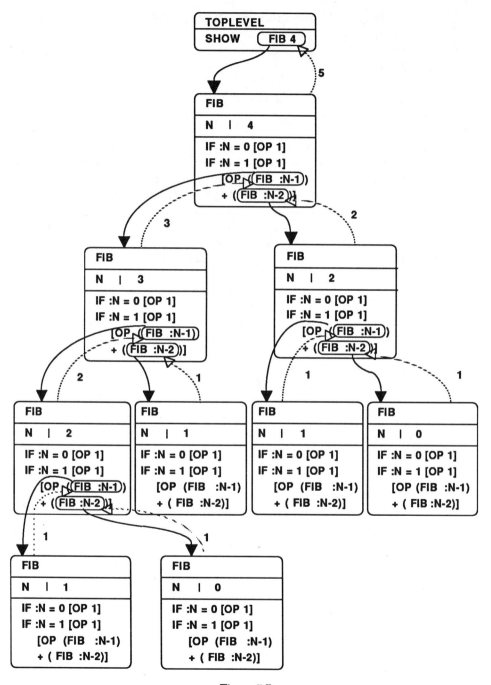

Figure 5.7

Tree Recursion

Example 5.5.2 Define a function named **ANYWHEREP** that requires two input values. The first input is assumed to be a word and the second is assumed to be a list. **ANYWHEREP** returns the word **TRUE** if the word appears anywhere inside the list and returns the word **FALSE** otherwise. For example,

ANYWHEREP "A [X D A L]	returns	TRUE
ANYWHEREP "A [[X Y] [A B C]]	returns	TRUE
ANYWHEREP "A [ ]	returns	FALSE
ANYWHEREP "A [D E [F G H]]	returns	FALSE

Note the difference between **ANYWHEREP** and **IS.MEMBERP** (See Exercise 4.5.5)

IS.MEMBERP "A [[X Y] [A B C]]	returns	FALSE
ANYWHEREP "A [[X Y] [A B C]]	returns	TRUE

The word **A** is not an element of the list **[[X Y] [A B C]]**, but it is part of the structure of the list.

Algorithm   The algorithm is a form of list recursion.

1. If the list is empty, then the result is **FALSE**.

2. If the first element of the list is equal to the first input to **ANYWHEREP**, the result is **TRUE**.

3. If the first element of the list is a word and (2) fails, then the result returned is found by seeing if the first input appears anywhere in the rest of the list.

4. If the first element of the list is in turn a list and the first input is anywhere in it, return **TRUE**.

5. Otherwise, the result is obtained by determining if the first input is anywhere in the rest of the list.

Logo Definition Transforming this algorithm into Logo, we have

```
TO ANYWHEREP :W :LIST
IF :LIST = [ ] [OP "FALSE]
IF :W = FIRST :LIST [OP "TRUE]
IF WORDP FIRST :LIST [OP ANYWHEREP :W BF :LIST]
IF ANYWHEREP :W FIRST :LIST [OP "TRUE]
   [OP ANYWHEREP :W BF :LIST]
END
```

This is the most complex algorithm yet, but it is worth spending time looking at the behavior of the evaluation. If you become confused about the meaning of the Logo definition, look back at the algorithm on the preceding page. After all, the program is just a direct transformation of it.

The final example of this section has the turtle draw a tree. It is interesting because the level-diagram representation of its execution looks like an upside-down version of the drawing it produces.

Example 5.5.3 Define a procedure named **TREE** that will cause the turtle to draw the picture of a tree on the screen.

Algorithm We must first define what the tree drawing should look like. Our view of a tree will be a primitive one, but the exercises offer some interesting variations. We will think of a tree as having a trunk and branches emanating from the trunk.

The trunk of the tree will be a vertical line segment whose length will be specified by an input to **TREE**. At the top of the trunk, the tree will branch out in two directions. These "branches" will, of course, be shorter than the trunk. At the end of each branch two more branches will sprout, each being somewhat shorter than its supporting branch. That is, we want our tree to look like the picture in Figure 5.8.

Figure 5.8

We need to specify more detail. We can think of the tree as a line segment with two smaller trees attached to one end of this line segment. The trunks of each of these smaller trees will point in directions that are, respectively, to the left and right of the main trunk. That is, we have developed the following scheme for drawing a tree. (We assume that the turtle has a heading of 0° initially.)

1. Move the turtle forward a specified amount.
(This is the trunk.)

2. Rotate the turtle left 45°. (We are getting ready to draw the tree growing to the left from the top of the trunk.)

3. Draw a tree from this position and heading but with a shorter trunk.
(We assume a trunk that is one-half the size of the supporting trunk.)

4. Rotate the turtle right 90° so that the turtle now has a heading of 45°.
(This assumes that Step 3 brought the turtle back to the same position and heading that it had before we drew the tree in 3.)

5. Draw a tree from this position and heading but with a trunk the same length as in Step 3.

6. Rotate the turtle left 45° and move it back the length of the trunk.
(The turtle is now back at its original position.)

Step 6 is extremely important. The assumption we made in (4) dictates that we must bring the turtle back to its original position. If you don't believe it, try defining **TREE** without it.

We made one oversight in the above algorithm. There is no terminating case. That is, we perform Steps 1, 2, and 3, which require we do 1, 2, and 3, which require we do . . . . The algorithm will never get around to doing (4), (5), and (6).

Since the turtle is asked to draw lines of smaller lengths, the solution is to stop drawing when the length of a line segment is so small we can no longer distinguish one segment from another. Let's decide that the terminating length will be 2. Thus, the first thing we need to do is check the length of the trunk. If it is less than 2, we stop.

The translation of our algorithm into Logo is:

```
TO TREE :TRUNKLENGTH
IF :TRUNKLENGTH < 2 [STOP]
FD :TRUNKLENGTH
LT 45
TREE :TRUNKLENGTH / 2
RT 90
TREE :TRUNKLENGTH / 2
LT 45
BACK :TRUNKLENGTH
END
```

Exercise 5.5.1. Describe the execution of **SHOW FIB 4** using the substitution and simplification model of computation.

† **Exercise 5.5.2.** Determine the total number of function calls to **FIB** required by the algorithm of Example 5.5.1 to compute fib(5), fib(6), and fib(n), where $n > 2$.

Exercise 5.5.3. Using the definition of **ANYWHEREP,** sketch the execution behavior of the following using level diagrams.

   a. **SHOW ANYWHEREP "A [ ]**
   b. **SHOW ANYWHEREP "A [[X Y] A [Q T]]**
† c. **SHOW ANYWHEREP "A [[X Y] W Z]**

Exercise 5.5.4. Using the definition of **ANYWHEREP,** describe the execution of the statements in Exercise 5.5.3 using the substitution and simplification model of computation.

Exercise 5.5.5. Define a function named **DELETE.ALL** that takes a list and a word as its two input values. It returns a copy of the list in which every occurrence of the word input in the list is deleted. For example,

   DELETE.ALL "A [A B [A B C]] returns **[B [B C]]**

Exercise 5.5.6. Draw the level diagram that describes the execution of **TREE 8**. Compare the level diagram you get with the actual picture of the tree produced by executing **TREE 8**.

Exercise 5.5.7. Modify the definition of **TREE** so that the angle becomes an input to the procedure. That is, **TREE 8 45**, where the length of the branch is 8 and the rotation angle is 45° will provide the same picture as **TREE 8** using the definition described earlier.

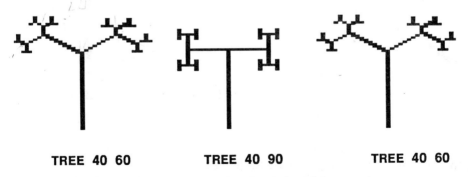

   TREE 40 60   **TREE 40 90**   **TREE 40 60**

† **Exercise 5.5.8.** Modify the original definition of **TREE** so that it will sprout three branches at the end of every supporting branch.

166 *Level Diagrams: Another Model For Describing Computation*

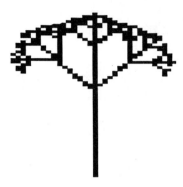

TREE 40

Exercise 5.5.9. Modify the definition of **TREE** in **Exercise 5.5.7** so that it contains a branching factor. That is, **TREE** will have three inputs: a branch length, an angle of rotation, and a number that indicates the depth of branching.

| TREE 40 5 45 | TREE 40 2 45 | TREE 40 4 60 |

5.6 Some Efficiency Considerations

The definitions of **AVERAGE**, a function that returns the average of a list of numbers (Section 4.5), and **FIB** (Section 5.5) have some inefficiencies in them. While efficiency should not be the primary consideration when defining an operation for the first time, at some point we might want to examine our definitions and eliminate extra computations in order to obtain the results much faster.

The definition for **AVERAGE**,

```
TO AVERAGE :LIST
OP (ADD :LIST) / (LENGTH :LIST)
END
```

requires Logo to examine the entire list of numbers twice. The first time, **ADD** examines the list in order to find the sum of all the numbers. Then, **LENGTH** does it again in order to count the number of elements in the list. We need both quantities, but it would be more efficient to do both at the same time.

Let's look at an example to see how we can compute the average of the numbers in the list **[3 5 7]** without examining each element of the list twice. We look at the first element of the list, add it to a running total (which is 0, initially) and increment a counter (initially, 0) by one. We repeat this step for each element in the list. When the list becomes empty, we divide the total by the counter. That is, we start out with three quantities: the list, a zero total, and a count of zero. Each is modified as follows:

LIST	TOTAL	COUNT
[3 5 7]	0	0
[5 7]	3	1
[7]	8	2
[]	15	3

We need to have three variables that describe the current state of the computation. We therefore define a function named **AVE1,** which has three input variables. The function **AVE1** has two cases to consider:

1. If the list is empty, then divide the total by the count.

2. If the list is not empty, call on **AVE1** with a new computational state.

    a. The new list input will be all but the first element of the current list.
    b. The total input will be increased by the first number in the current list.
    c. The count input will be increased by 1.

Items 1 and 2 outline the termination case and the general case for list recursion. Thus, the definition of **AVE1** looks like this:

```
TO AVE1 :LIST :TOTAL :COUNT
IF EMPTYP :LIST
   [OP :TOTAL / :COUNT]
   [OP AVE1 (BUTFIRST :LIST)
           (:TOTAL + FIRST :LIST)
           (:COUNT + 1)]
END
```

To get things off the ground, we need to call **AVE1** with values of zero for the total and count inputs. That is, to compute the average of the numbers in the list **[3 5 7]**, we must write

    AVE1 [3 5 7] 0 0

There is one displeasing aspect to this solution, namely, that we must now type in two additional values when we want to compute the average of a list of numbers. There is also a more serious problem. If we call **AVERAGE** with an empty list,

    AVE1 [ ] 0 0

it will blow up in our faces because it will immediately try to divide by zero. Both objections can be repaired by defining an additional function named **AVERAGE** that will return zero if the list is empty and supply the additional input values for **AVE1** if the list is not empty:

```
TO AVERAGE :LIST
IF EMPTYP :LIST [OP 0]
   [OP AVE1 :LIST 0 0]
END
```

To compute the average of the numbers in the list [3 5 7], we just type:

AVERAGE [3 5 7]

The inefficiency in the definition of **FIB** is more serious. We repeatedly compute the same value during the computation. For example, in **FIB 4** we compute **FIB 2** twice and **FIB 1** three times. Another way to think about computing the nth element of the Fibonacci sequence is to generate successive pairs of the sequence and to count up until we get to the one we want.

n	fib(n)	fib($n + 1$)
0	1	1
1	1	2
2	2	3
3	3	5
4	5	8
5	8	13
6	13	21
.	.	.
.	.	.
.	.	.

Looking at this table of pairs of Fibonacci numbers, we see that the nth Fibonacci number is the first component of the nth pair if we regard the zeroth pair to be the first two numbers in the Fibonacci sequence. That is, we can define **FIB** as

```
TO FIB :N
    OP FIB1 :N 1 1
END
```

where **FIB1 :N 1 1** returns the first component of the nth pair when each number in the zeroth pair is **1**.

The definition of **FIB1** is based on the following observations:

1. If n is greater than zero, the nth pair in the sequence of pairs starting with a particular pair is the same as the $(n - 1)$st pair in the sequence starting with the next pair.

2. The first component of any pair is the second component of the previous pair.

3. The second component of any pair is just the sum of the components of the previous pair.

We can formulate this idea in terms of counting recursion as follows:

> **Termination Case** If $n = 0$, then return the first component of the current pair.
>
> **General Case** If $n > 0$, then return the first component of the $(n - 1)$st pair starting with the next pair of Fibonacci numbers.

The Logo definition is:

```
TO FIB1 :N :P1 :P2
IF :N = 0 [OP :P1]
    [OP FIB1 :N-1 :P2 :P1+:P2]
END
```

This definition of **FIB** will be more efficient because we compute intermediate values in the sequence only once. For example, in generating the fourth pair, **FIB 2** is computed only once.

The level diagram describing the execution of **SHOW FIB 4** is given in Figure 5.9.

Some Efficiency Considerations 171

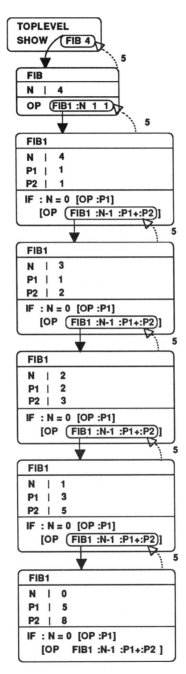

Figure 5.9

We no longer have a treelike computation. It is purely linear. The following exercises ask you to compare this version of **FIB** with the one presented earlier by experiencing it as well as by putting numeric values on the amount of work done by each. The difference in the amount of work done by each is very significant.

The primary concern in developing algorithms for solving problems is to find natural algorithms. At times efficient algorithms are important, but improvements in efficiency come more easily after a straightforward approach has been achieved. The more efficient algorithms for **FIB** and **AVERAGE** are not simpler nor are they more readable.

Exercise 5.6.1. Describe the execution of **AVERAGE [3 5 8 9]** using simplification and substitution and the definitions of **AVERAGE** and **AVE1** in this section.

Exercise 5.6.2. Describe the execution of **AVERAGE [3 5 8 9]** using level diagrams and the definitions of **AVERAGE** and **AVE1** in this section.

Exercise 5.6.3. What is the number of function calls needed to compute the average of a list of n elements using the definition of **AVERAGE** given in this Section? Using the definition in Section 4.5?

Exercise 5.6.4. Compute **FIB 20** using each of the following definitions.

a. TO FIB :N
   IF :N=0 [OP 1]
   IF :N=1 [OP 1]
     [OP (FIB :N-1) + (FIB :N-2)]
   END

b. TO FIB :N
   FIB1 :N 1 1
   END

   TO FIB1 :N :P1 :P2
   IF :N = 0 [OP :P1]
     [OP FIB1 :N-1 :P2 :P1+:P2]
   END

† **Exercise 5.6.5.** Determine the number of function calls necessary to compute **FIB 5**, **FIB 6**, and **FIB :N** for **:N > 2**, using each definition in Exercise 5.6.4.

† **Exercise 5.6.6.** The number of function calls performed is one measure of efficiency for an algorithm but not always the important measure. Compute the number of additions performed in evaluating **FIB 5**, **FIB 6**, and **FIB :N** for each definition in Exercise 5.6.4. How many additions are performed for each in the evaluation of **FIB 20**?

5.7 Tail Recursion

A form of recursion, where the last expression or statement executed before returning a call to itself, is called **tail recursion**. A complete definition follows. This type of recursion has important execution-time characteristics, which are examined next.

> The definition of a function or a procedure is said to be **tail recursive** if it contains at least one recursive call and the function or procedure returns immediately after each recursive call has completed.

An example of a function whose definition is tail recursive is that of **NTH** presented in Chapter 4:

```
TO NTH :LIST :N
IF (:N = 1)
   [OP FIRST :LIST]
   [OP NTH BUTFIRST :LIST :N-1]
END
```

The function **NTH** is tail recursive because whatever is computed by the recursive call, **NTH BUTFIRST :LIST :N-1**, is immediately returned.

For an example of a function whose definition is not tail recursive, recall the definition of **POWER**:

```
TO POWER :X :N
IF :N = 0 [OP 1] [OP :X * POWER :X :N-1]
END
```

After computing the value of the recursive call to **POWER**, we need to perform a multiplication before outputting a value.

Procedures may or may not be tail recursive. For example, consider the following definitions:

```
TO  DRAWSQUARES.ONE :DIST
IF :DIST > 0 [DRAWSQUARES.ONE :DIST-1]
FD :DIST
RT 90
END
```

```
TO  DRAWSQUARES.TWO :DIST
IF :DIST < 0 [STOP]
FD :DIST
RT 90
DRAWSQUARES.TWO :DIST-1
END
```

The definition of **DRAWSQUARES.TWO** is tail recursive because it returns immediately after performing the recursive call. However, **DRAWSQUARES.ONE** is not tail recursive because after performing the recursive call to itself, it must execute a **FD :DIST** and **RT 90** before returning.

The significance of tail recursion comes during the execution of a function or procedure call. If we execute **DRAWSQUARES.TWO 1000,** the turtle will draw a picture of an inward spiral (assuming that the boundary type of the graphics screen is **WINDOW**) and stop. However, executing **DRAWSQUARES.ONE 1000** will probably cause an error message before the turtle does anything. Using the substitution and simplification model, we can see why this might happen:

DRAWSQUARES.ONE 1000

    =>   IF 1000 > 0 [DRAWSQUARES.ONE 999]
         FD 1000
         RT 90

    =>   DRAWSQUARES.ONE 999
         FD 1000
         RT 90

    =>   IF 999 > 0 [DRAWSQUARES.ONE 998]
         FD 999
         RT 90
         FD 1000
         RT 90

    =>   DRAWSQUARES.ONE 998
         FD 999
         RT 90
         FD 1000
         RT 90

    =>   IF 998 > 0 [DRAWSQUARES.ONE 997]
         FD 998
         RT 90
         FD 999
         RT 90
         FD 1000
         RT 90

       .
       .
       .

No drawing will be done until the value of **:DIST** becomes negative; also, the Logo system is forced to remember all the **FD** and **RT** statements that are generated in each recursive call. Since the computer has a finite amount of memory, Logo could very easily run out before any drawing is performed.

Executing **DRAWSQUARES.TWO 1000** does not require any "remembering" because the **FD** and **RT** procedures are executed before the recursive call. That is,

    DRAWSQUARES.TWO 1000

        => IF 1000 < 0 [STOP]
            FD 1000
            RT 90
            DRAWSQUARES.TWO 1000-1

        => DRAWSQUARES.TWO 999

        => IF 999 < 0 [STOP]
            FD 999
            RT 90
            DRAWSQUARES.TWO 999-1

        => DRAWSQUARES.TWO 998

        .
        .
        .

There are no statements to be remembered during the execution of a tail-recursive function or procedure. Memory used during the computation becomes available for later computations. Thus, tail-recursive definitions should be preferred whenever there is a choice.

Another way to check for tail recursion is by looking at the level diagram for a recursively defined function or procedure. If each frame can be replaced by the lower-level frame it creates without changing the results, then it is tail recursive.

Exercise 5.7.1. Determine which of the following functions were given tail-recursive definitions. Explain your answer for each one.

    † a. **SUMSQ**   (Section 3.9)
    † b. **AVERAGE**   (Section 4.5)
       c. **ADD**   (Section 4.5)
    † d. **REVERSE**   (Section 4.5)
    † e. **BUTNTH**   (Section 5.4)

Exercise 5.7.2. Determine which of the following functions were given tail-recursive definitions. Explain your answer for each one. If any were not tail-recursive, could they be put in a tail-recursive form?

    a. **LENGTH**   (Exercise 4.5.1)
    b. **IS.MEMBERP**   (Exercise 4.5.5)
    c. **MEMBERN**   (Exercise 4.5.6)
    d. **ALL.BUT.FIRST**   (Exercise 5.4.3)
    e. **ALL.BUT**   (Exercise 5.4.5)

Exercise 5.7.3. Determine whether or not each of the following procedures is tail recursive. Explain your answer for each.

    a. **DOSQUARE**   (Section 3.7)
    b. **GROW**   (Section 3.7)
    c. **SPIRAL**   (Section 4.7)
    d. **SPIRAL.IN**   (Section 4.7)
    e. **RANDOMSPIRAL**   (Section 4.7)
    f. **CONNECT.ALL**   (Section 4.8)
    g. **CONNECT.POINTS** (Section 4.8)

Exercise 5.7.4. Discuss the execution of **DOSQUARE.ONE** and **DOSQUARE.TWO**, where these procedures are defined as follows:

```
TO DOSQUARE.ONE :DIST          TO DOSQUARE.TWO :DIST
FD :DIST                       DOSQUARE.TWO :DIST
RT 90                          FD :DIST
DOSQUARE.ONE :DIST             RT 90
END                            END
```

Chapter 6
Variables

6.1 Introduction

This chapter introduces the concepts of global variables and local variables that are not input variables. These additions provide flexibility in writing programs, but they have undesirable characteristics as well. They can cause unexpected results with no clue as to the source of the difficulty. They also cause serious problems for the models of computation that have been described in earlier chapters. In fact, the substitution and simplification model cannot cope with these additions at all.

The use of global and local variables in programming applications is illustrated in an example that deals with maintaining a directory of telephone numbers. The use of local variables is further illustrated in programs dealing with operations on sets. This chapter fully discusses the ideas these features encapsulate and their appropriate uses. Misuse and lack of understanding will eventually lead to programs that do not work or that contain mind-boggling bugs.

6.2 Global Variables

Global variables are variables that exist at the top level. They are useful in that they are permanent additions to the workspace and are always available at the top level. They are created by the **MAKE** procedure. **MAKE** requires two inputs: a word denoting the name of the variable and the value to be associated with that variable. For example,

    ?MAKE "SCORES [77 93 44 65 81]

creates a global variable, **:SCORES**, whose value is the list **[77 93 44 65 81]**. **MAKE** is completely described in the next section.

To see the value of the variable we simply ask for it to be shown:

?SHOW :SCORES
[77 93 44 65 81]

To compute the average of the numbers in the list we just call the **AVERAGE** function defined in Chapter 5 with **:SCORES** as the input expression:

?SHOW AVERAGE :SCORES
72

The value of a global variable can be modified by **MAKE** as well:

?MAKE "SCORES [10 20 30 40 50]
?SHOW AVERAGE :SCORES
30

We incorporate global variables into our level diagrams by including them in the toplevel frame. For example, after these statements have been executed, the toplevel frame is as in Figure 6.1.

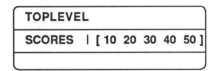

Figure 6.1

The execution of a **MAKE** statement causes the name of the variable and its value to be added to the workspace dictionary. Whenever a **SAVE** statement is executed, the global variables and their values are saved in the named file along with the definitions of all functions and procedures. The file will contain the **TO** forms of the functions and procedures and a **MAKE** statement for each global variable that exists when the **SAVE** is executed.

For example, if the workspace contains the definitions of **AVERAGE**, **ADD**, and **LENGTH** as well as the global variable **:SCORES**, then executing

    ?SAVE "STATS

causes the file named **STATS** to contain the following text:

    MAKE "SCORES [10 20 30 40 50]

    TO AVERAGE :LIST
    OP (ADD :LIST) / LENGTH :LIST
    END

    TO ADD :LIST
    IF EMPTYP :LIST [OP 0] [OP (FIRST :LIST) + ADD BF :LIST]
    END

    TO LENGTH :LIST
    IF EMPTYP :LIST [OP 0] [OP 1 + LENGTH BF :LIST]
    END

When the file is **LOAD**ed back into the workspace, the global variable **:SCORES** and the saved value are restored to the workspace along with the definitions of **AVERAGE, ADD** and **LENGTH**.

A global variable remains in the dictionary until it is removed by a procedure named **ERN** (erase name). Erasing the variable **:SCORES** and then trying to display it results in a familiar error message:

    ?ERN "SCORES
    ?SHOW :SCORES
    :SCORES HAS NO VALUE

The inputs to **MAKE** can be any Logo expressions, as long as the evaluation of the first input expression is a word that names a variable. For example, consider the following interaction:

    ?MAKE "X [A B C D]
    ?MAKE FIRST [SALLY JANE] 357
    ?MAKE FIRST :X "REX
    ?MAKE :A FIRST BF [10 20 30 40]
    ?SHOW :X

```
[A B C D]
?SHOW :SALLY
357
?SHOW :A
REX
?SHOW :REX
20
```

The first **MAKE** statement creates the global variable named **X** and gives it the list **[A B C D]** as its value. The next creates the variable named **SALLY** (the value of **FIRST [SALLY JANE]**) with the number **357** as its value. The third creates a variable named **A** (the value of **FIRST :X**) and gives it the word **REX** as its value. The last **MAKE** statement creates the variable named **REX** (the value of **:A**) and gives it the number **20** as its value.

The execution of these toplevel statements is pictured (Figure 6.2) as a sequence of toplevel frames indicating the changing state. This sequence of toplevel frames is called a **state diagram**. Every time a variable is modified, a new level diagram is created by copying the current one with the modification.

State (i) shows there are no global variables at the time we begin the execution of **MAKE "X [A B C D]**. State (ii) shows the state when **MAKE FIRST [SALLY JANE] 357** is executed. States (v) and (vi) represent the states before and after executing **MAKE "X BF :X**.

182 *Variables*

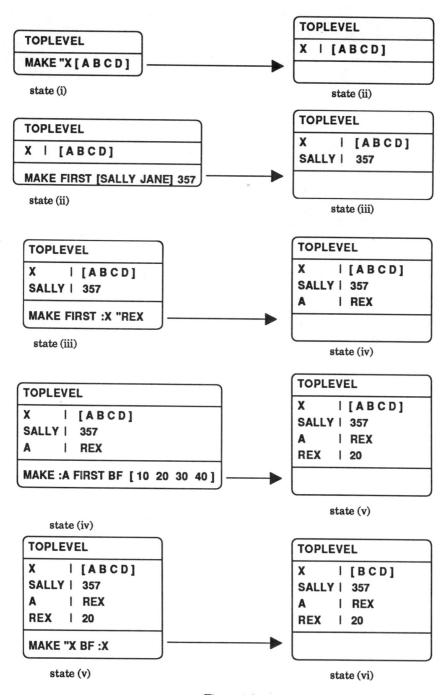

Figure 6.2

Exercise 6.2.1. Assuming that there are no global variables initially, draw the state diagram that corresponds to the following sequence of statements executed at the toplevel:

```
MAKE "NAMES [ZELDA SAM NAME]
MAKE LAST :NAMES BUTLAST :NAMES
MAKE LAST :NAME FIRST BF :NAMES
MAKE WORD LAST :NAMES "S BF :NAME
```

6.3 The Dynamic Scoping Rule

To fully understand the behavior of executing a **MAKE** statement we must first understand how a reference to a variable is resolved. We begin with an example that shows how to define a function that accesses the value of a variable that is not an input variable.

Example 6.3.1 If we execute

```
?MAKE "NUMLIST [10 20 30]
```

and define functions named **ADD.NUMS** and **ADD** as

```
TO ADD.NUMS
OP ADD :NUMLIST
END

TO ADD :LIST
IF EMPTYP :LIST [OP 0]
   [OP (FIRST :LIST) + ADD BF :LIST]
END
```

we can display the sum of the numbers in the list stored in **:NUMLIST** by executing **SHOW ADD.NUMS**:

```
?SHOW ADD.NUMS
60
```

The first two frames of the level diagram describing the execution of **SHOW ADD.NUMS** is given in Figure 6.3 (a). The level-one frame says to compute **ADD :NUMLIST** but there is no value for the variable **:NUMLIST** in that frame. Since there is a variable **:NUMLIST** in the level-zero frame, the value used as the input to **ADD** will be the value contained in the level-zero frame (the list **[10 20 30]**). The computation will proceed as described in Figure 6.3 (b).

The convention used in resolving a reference to a variable is called the dynamic scoping rule and is stated next.

Dynamic Scoping Rule
The reference to a variable is the first binding found when searching back through the sequence of creating frames starting with the frame containing the reference, and ending with the toplevel frame. If no binding is found in any of these frames the variable is undefined.

That is, first look for the variable in the referencing frame. If no binding for the variable is there, look for a binding in the frame that created the referencing frame, then in the frame that created that frame, and so on. The search is terminated when a binding is found, or the toplevel frame is reached.

The Dynamic Scoping Rule 185

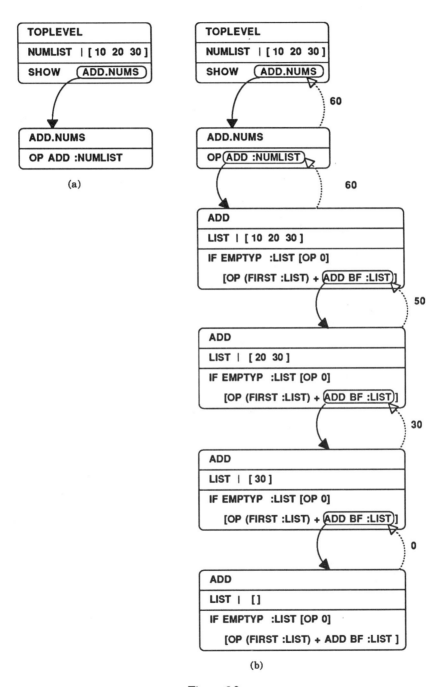

Figure 6.3

Example 6.3.2 shows that access to a global variable can be blocked by an intermediate computation.

Example 6.3.2 Suppose that the definition of a procedure named DISPLAY.SUMS is given by

```
TO DISPLAY.SUMS :NUMLIST
SHOW ADD.NUMS
SHOW ADD :NUMLIST
END
```

and that we execute DISPLAY.SUMS [1 2 3] at the toplevel with the global variable :NUMLIST bound to [10 20 30]. That is,

```
?DISPLAY.SUMS :NUMLIST
60
60
```

Figure 6.4 shows part of the level diagram that describes this computation. At level one we first execute ADD.NUMS, which causes the creation of the leftmost level-two frame.

Before we call ADD we must resolve the reference to the variable :NUMLIST. Since it is not found in the level-two frame we look at the preceding frame. Since there is a binding for :NUMLIST we call ADD with the list [1 2 3] as its input.

After computing the sum (6), it is returned to level-one frame where it is displayed. The reference to :NUMLIST in the statement SHOW ADD :NUMLIST is resolved by using the value contained in the level-one frame. The value of ADD :NUMLIST gives 6 again. Thus, the number 6 is displayed again.

Note that there are two variables named :NUMLIST while DISPLAY.SUMS is being executed, and that the level-one occurrence of :NUMLIST effectively blocks all access to the level-zero occurrence of :NUMLIST.

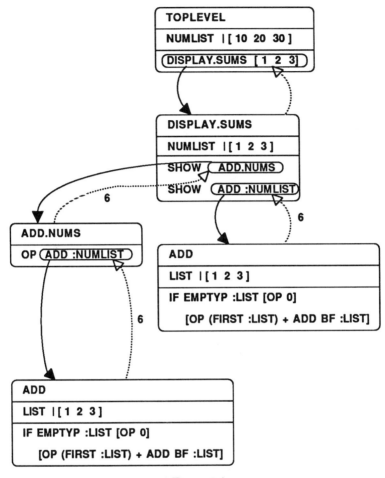

Figure 6.4

The best way to avoid the problems of global variables is not to use them. However, they do serve a useful purpose in programming applications in that they allow data to be kept around for easy access. The telephone directory application discussed in the next section is an example.

In general, global variables should be used *very* sparingly. When they are used, you need to remember their names and never use those names for input variables in any functions or procedures that reference the global variables. Not remembering to do this usually results in bugs mysteriously appearing with no clue as to their origins. This is especially true when large programs are written and expanded over a period of several days or longer, after you have had plenty of time to forget their names.

† **Exercise 6.3.1.** Suppose that we replace the definition of **DISPLAY.SUMS** in Example 6.3.2 by

```
TO DISPLAY.SUMS :LIST
SHOW ADD.NUMS
SHOW ADD :LIST
END
```

What happens when we execute **DISPLAY.SUMS [1 2 3]**? Explain why the effect differs from the example.

† **Exercise 6.3.2.** The **GRAPH** program in Section 4.9 calls **GENERATE.POINTS** with three inputs:

```
TO GRAPH :LEFTEND :RIGHTEND :INCREMENT
CONNECT.POINTS  GENERATE.POINTS :LEFTEND
                                :RIGHTEND
                                :INCREMENT
END
```

(a) Redefine **GENERATE.POINTS** so that it has only one input (**:LEFTEND**). That is, the definition of **GRAPH** will look like this:

```
TO GRAPH :LEFTEND :RIGHTEND :INCREMENT
CONNECT.POINTS  GENERATE.POINTS :LEFTEND
END
```

(b) Can you redefine **GENERATE.POINTS** so that no inputs are required?

6.4 The Make Statement

In Section 6.2 we examined the use of **MAKE** to create and modify global variables. **MAKE** may also be used in a statement contained in the body of a procedure but its behavior is not limited to creating and modifying global variables. The complete description of **MAKE** and its effect is given in the next definition. Following the definition we illustrate its behavior with three examples.

> **MAKE** *word object*
> **MAKE** is a binary procedure. The variable named by the first input value is found using the dynamic scoping rule and its value is set to *object*. If no such variable is found, a global variable by that name is created and set to *object*.

Therefore, **MAKE** can create global variables or modify any variable (including a global one) in a currently active computation. Example 6.4.1 illustrates the use of **MAKE** in creating and assigning a value to a global variable.

Example 6.4.1 Consider the following procedure definition:

    TO ADD.GLOBAL :LIST
    MAKE "TOTAL ADD :LIST
    END

Note that when we execute **ADD.GLOBAL [1 2 3 4]**, no value is returned. This marks the difference between procedures and functions:

> We call a procedure to cause a side effect.
> We call a function to return a value.

Assuming that there are no global variables, the side effect of executing **ADD.GLOBAL [1 2 3]** is to create a global variable, **:TOTAL** whose value is 6.

The description of the execution of **ADD.GLOBAL [1 2 3]** is given by the state diagram in Figure 6.5. The list **[1 2 3]** is passed to **ADD.GLOBAL** as usual. Next, **ADD [1 2 3]** is computed, and the value (6) is returned to the level-one frame. Then the **MAKE** statement is executed. Since no variable named **TOTAL** exists, the global variable, **:TOTAL** is created and set to 6.

A new level diagram is created indicating that a change in state has occurred. The arrow indicating a change of state has occurred also marks the line that caused the state change. Execution then picks up in state (ii). Since there is nothing left to do in **ADD.GLOBAL** the level-one frame returns, leaving the global variable, **TOTAL**, with a value of 6 (final state).

Subsequent execution of **ADD.GLOBAL** at the toplevel will result in the modification of **:TOTAL**. For example,

>?ADD.GLOBAL [10 20 30]
>?SHOW :TOTAL
>**60**

The Make Statement

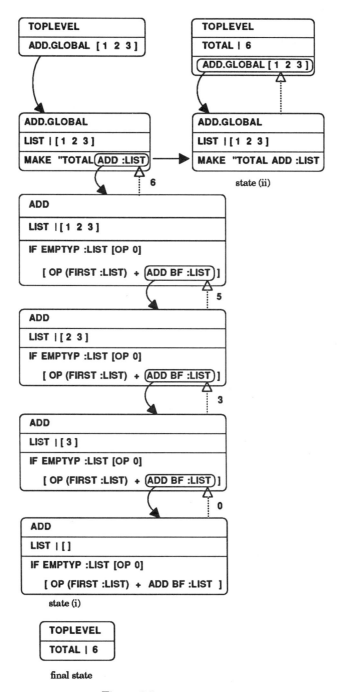

Figure 6.5

The next example illustrates a common programming error in conjunction with the use of the **MAKE** statement.

Example 6.4.2 Suppose that a procedure named **ADD.GLOBAL1** is defined as follows:

```
TO ADD.GLOBAL1 :TOTAL
MAKE "TOTAL ADD :TOTAL
END
```

It is identical to **ADD.GLOBAL** except that the input variable is named **TOTAL** rather than **LIST**. However, if we execute **ADD.GLOBAL1 [1 2 3]** (assuming that no global variables exist), the effect is very different. Figure 6.6 describes this computation.

After computing the sum, the **MAKE** statement is executed. Using the dynamic scoping rule, Logo determines that the level-one variable **:TOTAL** is to be modified. That is, no global variable is created because a variable by that indicated name already exists. After the execution of **MAKE** (but before the return of **ADD.GLOBAL1**), we are left with state (ii). Note that the original value of **:TOTAL** (the list **[1 2 3]**) is destroyed. After returning from the level-one call to **ADD.GLOBAL1,** we are left with the same state we started with. We went through all the work of computing the sum of a list of numbers but have nothing to show for it.

The Make Statement

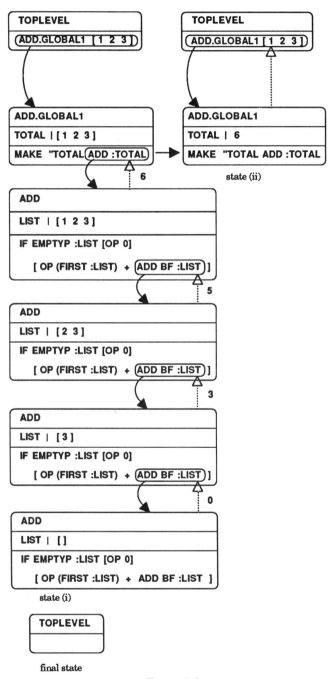

Figure 6.6

The **ADD.GLOBAL1** example illustrates a problem that is often encountered with the use of global variables. If an input variable of a function or procedure has the same name as a global variable, then access to the global variable by the same name is blocked whenever that function or procedure is active. All references to that variable in the function or procedure are to the input variable. All references to the global variable by functions or procedures called at a lower level are blocked as well.

Example 6.4.3 Suppose we have a list containing the names **JOHN, MARY**, and **ZELDA**, and that we store this list in a global variable **:NAMES**. We can do this by entering the following statement:

    ?MAKE "NAMES [JOHN MARY ZELDA]

Now define a procedure that takes a list of names as its input and adds these names to the list:

    TO ADD.NAMES :NAMELIST
        MAKE "NAMES SENTENCE :NAMELIST :NAMES
    END

To add **SAM** and **SALLY** to the list, we simply type in:

    ADD.NAMES [SAM SALLY]

Try it. The following interaction indicates that all is well:

    ?SHOW :NAMES
    [JOHN MARY ZELDA]
    ?ADD.NAMES [SAM SALLY]
    ?SHOW :NAMES
    [SAM SALLY JOHN MARY ZELDA]

Now, define another procedure named **ADD.MORE** that takes a list of names as its input and calls **ADD.NAMES** to add them to the global list:

    TO ADD.MORE :NAMES
        ADD.NAMES :NAMES
    END

So far, so good: **ADD.MORE** appears to make the computation more complicated, but it does not appear that it should hurt anything. Actually, using **ADD.MORE** causes the desired effect to fail. Consider the following interaction:

```
?SHOW :NAMES
[SAM SALLY JOHN MARY ZELDA]
?ADD.MORE [ASTER TOM]
?SHOW :NAMES
[SAM SALLY JOHN MARY ZELDA]
```

ASTER and **TOM** seemed to get lost somewhere along the line. We didn't even get an error message to provide a clue to what went wrong.

We can trace down the problem by describing the execution of **ADD.MORE [ASTER TOM]** using level diagrams and remembering the dynamic scoping rule. See Figure 6.7.

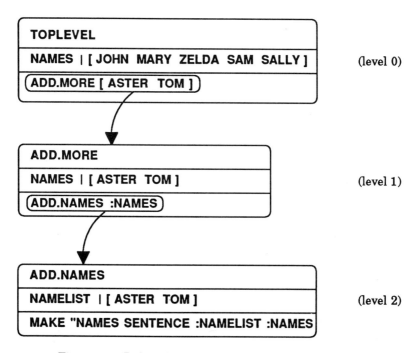

Figure 6.7 Before the Execution of the Body of **(ADD.NAMES)**

The toplevel frame calls **ADD.MORE** with the list **[ASTER TOM]** as its input value. **ADD.MORE** then calls **ADD.NAMES**. The input value to **ADD.NAMES** is the value of **:NAMES**. Using the scoping rule, the value of **:NAMES** is the list **[ASTER TOM]**. We arrive in the level-two frame in fine shape.

Inside **ADD.NAMES**, Logo discovers that it must simplify **SENTENCE :NAMELIST :NAMES**. The value of **:NAMELIST** is found in the same frame, the list **[ASTER TOM]**. The value of **:NAMES** is not found in the level-two frame, and the scoping rule says to look in the previous frame. Here, we find a value for **:NAMES**, the list **[ASTER TOM]**. This is the beginning of our problem. That is, the simplification of **SENTENCE :NAMELIST :NAMES** proceeds as follows:

    SENTENCE :NAMELIST :NAMES
    => SENTENCE [ASTER TOM] [ASTER TOM]
    => [ASTER TOM ASTER TOM]

This is not the value we intended to give to the global variable, but the problem is made even more mysterious by the execution of **MAKE "NAMES [ASTER TOM ASTER TOM]**.

Again, the variable name of **:NAMES** is not found in the level-two frame. In this case, the scoping rule says to look in the level-one frame. We find **NAMES** in the level-one frame, so **MAKE** changes its value. That is, just before the level-two frame returns, the state of the computation is described by the level diagram in Figure 6.8.

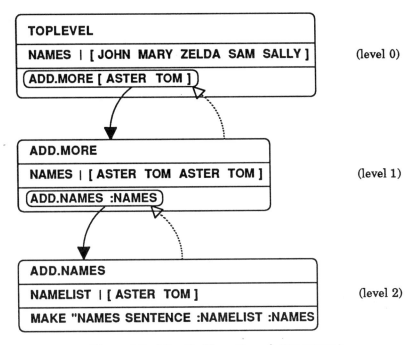

Figure 6.8 After the Execution of (ADD.NAMES)

Now, **ADD.NAMES** returns to the level-one frame. Since this completes the computation in the level-one frame as well, we arrive back at the toplevel with no change in the global variable **:NAMES**.

The **MAKE** statement cannot be incorporated into the simplification model of computation. It doesn't provide a mechanism to keep track of the state (variable bindings) of a computation or the modification of a variable. To see this, consider the simplification of **ADD.GLOBAL [1 2 3]** where **ADD.GLOBAL** is defined in Example 6.4.1:

```
ADD.GLOBAL [1 2 3]
=>  MAKE "TOTAL ADD [1 2 3]
=>  MAKE "TOTAL 6
```

MAKE does not return a value. Thus, we cannot simplify this last statement to 6. Also note that the simplification of **ADD.GLOBAL1 [1 2 3]** (Example 6.4.2) proceeds in the same way

    ADD.GLOBAL1 [1 2 3]
    => **MAKE "TOTAL ADD [1 2 3]**
    => **MAKE "TOTAL 6**

but the effect of these computations is very different.

The simplification model of computation provides a useful tool in understanding the effects of programs that do not use **MAKE** statements or that refer to variables other than its own input variables. It provides an easily understood set of rules that describe how a computation is carried out for a substantial part of the Logo language. We should not discard the model simply because it does not model the entire language. In fact, one could argue that the language should be changed to conform to the model, since the inclusion of **MAKE** does not add any computational power to the language.

† **Exercise 6.4.1.** Using the following definition of **MAKE.TOTAL**,

    TO MAKE.TOTAL :TOTAL
    MAKE "TOTAL ADD :TOTAL
    END

describe the state diagram for the execution of the following sequence of statements:

    ?MAKE "TOTAL 0
    ?MAKE.TOTAL [1 2 3]

Exercise 6.4.2. Draw the state diagram for the execution of **ADD.NAMELIST [ASTER TOM]** after executing the statement

    ?MAKE "NAMES [SALLY BILL]

where **ADD.NAMELIST** is defined by

> **TO ADD.NAMELIST :NAMELIST**
> **ADD.NAMES :NAMELIST**
> **END**

The definition of **ADD.NAMES** is the one in Example 6.4.3. What is the value of the global variable **:NAMES** after the execution of **ADD.NAMELIST [ASTER TOM]**?

† **Exercise 6.4.3.** Draw the state diagram for the execution of **ADD.MORE.NAMES [ASTER TOM]** in the context of the following interaction:

> **?MAKE "NAMES [SALLY BILL]**
> **?ADD.MORE.NAMES [ASTER TOM]**

The definition of **ADD.MORE.NAMES** is

> **TO ADD.MORE.NAMES :NAMES**
> **IF EMPTYP :NAMES [STOP]**
>   **[ADD.ONE.NAME FIRST :NAMES**
> **ADD.MORE.NAMES BUTFIRST :NAMES]**
> **END**

and the definition of **ADD.ONE.NAME** is

> **TO ADD.ONE.NAME :NAMELIST**
> **MAKE "NAMES FPUT :NAMELIST :NAMES**
> **END**

What happens when **ADD.MORE.NAMES [ASTER TOM]** is executed? How can the problem be fixed?

Exercise 6.4.4. Describe the state diagram for the execution of the sequence

> **MAKE "NAMES [JOHN]**
> **ADD.MORE [ZELDA SAM]**
> **ADD.MORE [SALLY BILL]**

where **ADD.MORE** is defined in Example 6.4.3.

200 Variables

Exercise 6.4.5. Draw the state diagram describing the execution of SHOW ADDEM [4 5 6], assuming the existence of a global variable named NUMS whose value is [1 2 3], where ADDEM is defined by

    TO ADDEM :NUMS
    OP ADD.LIST
    END

    TO ADD.LIST
    OP ADD :NUMS
    END

Exercise 6.4.6. Draw the state diagram describing the execution of the following sequence of statements:

    ?MAKE "X "HELLO
    ?ALTER :X
    ?GREET :X
    ?SHOW SE :X :Y

where ALTER and GREET are defined as follows:

    TO ALTER :Y
    MAKE "X LIST :X :Y
    SHOW :X
    END

    TO GREET :X
    MAKE "Y :X
    SHOW :X
    ALTER :Y
    SHOW :X
    END

6.5 Telephone Directory Program

We want to develop and maintain a telephone directory of our acquaintances. We want to create the directory and store it permanently.

We want to be able to look up the phone number of a particular individual, add new entries to it, delete entries, or modify an entry. This is a particular example of a more general problem called **symbol-table manipulation**.

A **symbol table** is a collection of name-value pairs. A telephone directory contains the names of people. The value associated with each name is the phone number for that person. Operations associated with a symbol table include adding entries, deleting entries, and modifying entries.

We can represent a symbol table as a list of name-value pairs. Each element of the symbol table is a list whose first element is a name and whose second element is the value corresponding to that name. For example, we could represent the telephone directory by the following list:

```
[[JOHN.DOE 555-1234]
 [SALLY.JONES 555-5678]
 [ZELDA.SMITH 555-2468]
 .
 .
 [MABEL.SMITH 555-1357] ]
```

This is the kind of thing we want to keep around all the time. That is, it is a data base that we will continually search and modify. One way to accomplish this task is to create a global variable whose value is this list. Any time we want to reference the list, we simply get the value bound to this variable. We can do this by entering a statement like the following:

?MAKE "DIR [[JOHN.DOE 555\-1234] [SALLY.JONES 555\-5678]]

The backslash character, \, is entered by typing CONTROL-Q and is used to tell Logo that we want the next character (in this case, -) to be treated as a normal character in a word.

To add an entry to the telephone directory, we might try the following expression:

FPUT [MABEL.SMITH 555\-1357] :DIR

This is not right, however, because **FPUT** is a function. That is, it does not change the value of **:DIR**, It returns a copy of the old list with the new entry added. We have to use **MAKE** to change the value of **:DIR**. This expression does compute the value we want :DIR to have, so the following statement does the trick:

    ?MAKE "DIR FPUT [MABEL.SMITH 555\-1357] :DIR

Although this last statement does the job, too much detail is specified. We want a procedure that takes a name and a phone number as inputs and does all the rest for us:

    TO ADD.ENTRY :NAME :VALUE
    MAKE "DIR FPUT LIST :NAME :VALUE :DIR
    END

To add the new entry to the directory, we enter the following statement:

    ?ADD.ENTRY "MABEL.SMITH "555\-1357

We can combine the creation of the telephone directory together with the function that adds a new entry. That is, there is no need to create the global variable :DIR as we just did. It can be created when the first entry is made by redefining ADD.ENTRY as follows:

    TO ADD.ENTRY :NAME :NUMBER
    IF NAMEP "DIR
      [MAKE "DIR FPUT LIST :NAME :NUMBER :DIR]
      [MAKE "DIR LIST LIST :NAME :NUMBER]
    END

We have introduced a new Logo primitive. It has a companion, so we describe both:

> **NAMEP** is a unary function whose input must be a word. It returns **TRUE** if its input is the name of an existing variable and **FALSE** otherwise.
>
> **THING** is a unary function whose input must be a word representing the name of a variable. It returns the value bound to the named variable.

The new **ADD.ENTRY** procedure checks to see if **DIR** is the name of an existing variable. If it is, the new entry is added to the directory. Otherwise, a directory with one entry is created.

To access the phone number of a particular person, we need to define a function that searches for the entry with the desired name and returns the phone number of that party. We name this function **FIND**. **FIND** has the name of a person as its single input and returns the phone number associated with that name. The definition of **FIND** is

```
TO FIND :NAME
OP LOOKUP :NAME :DIR
END
```

LOOKUP is a function that requires two inputs, a name and a symbol table. It does the actual searching and returns the phone number if it is found in the table. Otherwise, it returns the word **NOT.FOUND**. The algorithm for **LOOKUP** is:

1. If the symbol table is empty, return the word **NOT.FOUND**.

2. If the name input corresponds with the name in the first entry, return the value associated with that name.

3. Otherwise, look up the name in the remaining entries of the symbol table.

The definition of **LOOKUP** is

```
TO LOOKUP :NAME :SYMTAB
IF EMPTYP :SYMTAB [OP "NOT.FOUND]
IF :NAME = FIRST FIRST :SYMTAB
   [OP LAST FIRST :SYMTAB]
   [OP LOOKUP :NAME BUTFIRST :SYMTAB]
END
```

The level diagram describing the execution of **SHOW FIND "SALLY. JONES** is given in Figure 6.9.

204 *Variables*

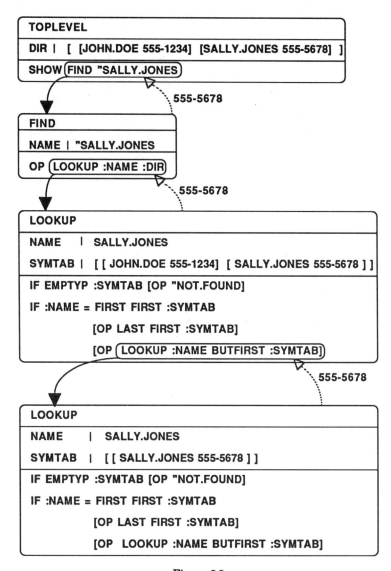

Figure 6.9

After binding the word **SALLY.JONES** TO **:NAME**, we begin executing the body of **FIND** in the level-one frame. It says to return the result obtained from **LOOKUP**. The name that we pass to **LOOKUP** is

SALLY.JONES, and the symbol table we specify is the telephone directory associated with the global variable :DIR. The value of :DIR is found using the dynamic scoping rule.

LOOKUP discovers that SALLY.JONES is not the first entry of the symbol table, so it tries looking up the name in the rest of the symbol table. The level-three frame discovers that the name matches with the first entry, and returns the corresponding phone number to the level-two frame. The level-two frame then returns the phone number of Sally Jones to the level-one frame, which returns the value to the toplevel frame.

Note that LOOKUP does not modify the value of :DIR. LOOKUP is given a copy of the value of the global variable :DIR.

Exercise 6.5.1. Redefine LOOKUP so that its second input is named DIR. Will this make any difference in the computation of FIND "SALLY.JONES?

† Exercise 6.5.2. Define a procedure named DELETE that takes a name as its input and deletes the entry in the directory corresponding to that name.

Exercise 6.5.3. Define a procedure named UPDATE that takes a name and a phone number as its inputs. If an entry with the name input already exists in the directory, then the entry is replaced by an entry with the new phone number. If such an entry does not currently exist, then the new entry is added to the directory.

† Exercise 6.5.4. Define a binary function named FIND.ALL that takes the initials of the first and last name of a person as its inputs and returns a list of names and numbers of all people with the specified initials. For example, FIND.ALL "S "J returns [[SALLY.JONES 555-5678] [SAM.JOHNSON 555-2468]] if these are the only people in the directory that have the initials S. J.

6.6 An Interactive Telephone Directory Program

Defining the telephone directory and the procedures for adding new entries, deleting entries, and updating entries provides the functionality portion of the application. What we want to do next is to combine these procedures into an interactive program that is convenient for the user. In applications programming, this part of the application is referred to as the *user interface.*

To make the basic functions and procedures easy to use, we display instructions telling the user how to use the application and have the user press a single key to indicate the operation (**LOOKUP**, **DELETE**, or **UPDATE** an entry) he or she wishes to perform.

That is, the program displays the instructions, reads a character indicating the desired procedure, requests the information needed to perform that procedure, and then calls one of our previously defined functions and procedures to do the work. Furthermore, the program should keep running to allow additonal requests.

The temptation to use another global variable is pretty high. That is, if we write the program just the way we described it, we would come up with something like the following:

```
TO TELEPHONE1
DISPLAY.DIRECTORY.INSTRUCTIONS
MAKE "COMMAND GET.COMMAND
IF :COMMAND = "F [SHOW FIND READNAME]
IF :COMMAND = "D [DELETE READNAME]
IF :COMMAND = "U [UPDATE READNAME READNUMBER]
IF :COMMAND = "Q [STOP]
TELEPHONE1
END
```

TELEPHONE1 calls **DISPLAY.DIRECTORY.INSTRUCTIONS** to display the instructions on the text screen. It then calls **GET.COMMAND** to read in a character from the keyboard. A global variable named **COMMAND** is created to save the command character.

The next four statements examine the command character. If the command character matches the character in one of the equality tests, the corresponding procedure is carried out after gathering the necessary information from the user.

READNAME is a function that gets a name from the user, and **READNUMBER** gets a phone number from the user. The name of a person is required to look up a phone number or to delete an entry. Both a name and a number are required to update an entry. If a **Q** is typed the program is terminated. Otherwise, the whole process is repeated by the recursive call to **TELEPHONE1**.

The **TELEPHONE1** program illustrates an inappropriate use of a global variable. The variable **:COMMAND** is used as a slot to hold onto the character entered by the user. It is used locally and it should be treated like a local variable. In keeping with our principle that global variables should be used only when appropriate, the next section presents two ways to make **:COMMAND** into a local variable.

† **Exercise 6.6.1.** Complete the **TELEPHONE1** program by defining the following functions and procedures:

    a. **DISPLAY.DIRECTORY.INSTRUCTIONS**
    b. **GET.COMMAND**
    c. **READNAME**
    d. **READNUMBER**

Exercise 6.6.2. Test the **TELEPHONE1** program for the following sequence of requests (assuming an initial directory that is empty):

    Add Mabel Smith, 555-1357, to the directory.
    Add John Doe, 555-1234, to the directory.
    Add Sally Jones, 555-2468, to the directory.
    Find the phone number of John Doe.
    Delete John Doe from the directory.
    Update Sally Jones phone number to 555-6789.

Exercise 6.6.3. Redefine **TELEPHONE1** so that the entire directory will be displayed nicely on the screen when the user presses the **A** key. The directory should be displayed 10 entries at a time, each entry on a different line. After 10 lines are displayed, the program should wait until the user presses a key indicating he or she is ready for the next 10 entries.

6.7 Local Variables

The previous section described the use of global variables as a way to keep information around for later use. Global variables should be used very sparingly, since they can cause lots of problems. You will be tempted to use them (especially if you have programmed in BASIC, FORTRAN, or Pascal) more frequently than is recommended. This section describes two ways to deal with the temptation.

In the **TELEPHONE1** program, we noted that the global variable, **:COMMAND**, should be treated as a local variable. One way to do this is to define a higher-level procedure to supply the command character and making **:COMMAND** an input variable to a lower-level procedure. By making **:COMMAND** an input variable, it disappears when the program is terminated. No toplevel junk is created. Here is the new program:

```
TO TELEPHONE2
DISPLAY.DIRECTORY.INSTRUCTIONS
TELEPHONE3 GET.COMMAND
END

TO TELEPHONE3 :COMMAND
IF :COMMAND = "F [SHOW FIND READNAME]
IF :COMMAND = "D [DELETE READNAME]
IF :COMMAND = "U [UPDATE READNAME READNUMBER]
IF :COMMAND = "Q [STOP]
TELEPHONE2
END
```

This version avoids the use of a global variable, but the program has become more complex. It takes more thinking and planning to come up with this solution.

The second alternative to the global variable version is to use a local variable that is not an input variable. That is, we would like the toplevel procedure, named **TELEPHONE**, to have a local variable that is not an input variable. It will be used temporarily to save the command character but it will disappear completely when **TELEPHONE** returns. Therefore, we want it to act like an input variable. In Logo, this is accomplished by using the **LOCAL** procedure.

> **LOCAL** word ... word
> **LOCAL** is a procedure with an arbitrary number of inputs, each of which must be a word. **LOCAL** adds variables named by the input words to the frame in which it appears. No values are associated with these variables when they are created.

Variables come in two flavors: global variables and local variables. Local variables are either input variables or variables created by **LOCAL**. We can make use of **LOCAL** in the telephone directory application as follows:

```
TO TELEPHONE
LOCAL "COMMAND
DISPLAY.DIRECTORY.INSTRUCTIONS
MAKE "COMMAND GET.COMMAND
IF :COMMAND = "F [SHOW FIND READNAME]
IF :COMMAND = "D [DELETE READNAME]
IF :COMMAND = "U [UPDATE READNAME READNUMBER]
IF :COMMAND = "Q [STOP]
TELEPHONE
END
```

The level diagram for the execution of **TELEPHONE** is given in Figure 6.10. The local variable, **:COMMAND**, is created in the **TELEPHONE** frame when the **LOCAL** statement is executed. It is given a value when the **MAKE** statement is executed. Like input variables, it disappears when the frame to which it belongs returns.

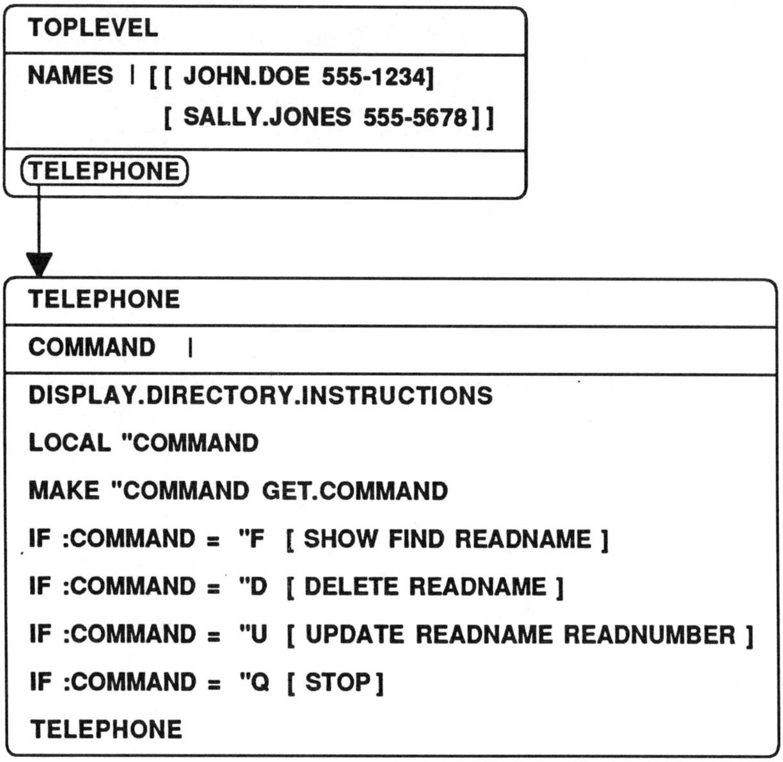

210 Variables

```
┌─────────────────────────────────────────────┐
│ TOPLEVEL                                    │
├─────────────────────────────────────────────┤
│ NAMES  |  [ [ JOHN.DOE 555-1234]            │
│           [ SALLY.JONES 555-5678 ] ]        │
│ (TELEPHONE)                                 │
└─────────────────────────────────────────────┘
              │
              ▼
┌─────────────────────────────────────────────┐
│ TELEPHONE                                   │
├─────────────────────────────────────────────┤
│ COMMAND    |                                │
│ DISPLAY.DIRECTORY.INSTRUCTIONS              │
│ LOCAL "COMMAND                              │
│ MAKE "COMMAND GET.COMMAND                   │
│ IF :COMMAND = "F  [ SHOW FIND READNAME ]    │
│ IF :COMMAND = "D  [ DELETE READNAME ]       │
│ IF :COMMAND = "U  [ UPDATE READNAME READNUMBER ] │
│ IF :COMMAND = "Q  [ STOP ]                  │
│ TELEPHONE                                   │
└─────────────────────────────────────────────┘
```

Figure 6.10

6.8 Procedural vs. Functional Programming

The previous sections introduced the idea of changing states by using **MAKE** to modify the value of a variable. When the modification is performed, the old value is lost forever. During the course of a computation, the value of a variable may change many times. The idea behind this style of programming is that at the time the computation has completed, the value (or values) associated with a certain variable (or variables) is (are) the result(s) we want from the computation. This style of programming is called **procedural programming** and is reflected in almost all programming languages.

Prior to the introduction of **MAKE**, all computations that produced "results" were performed by defining operations that simply returned the desired result. This style of programming is called **functional programming**. Example 6.8.1 illustrates the difference.

Example 6.8.1 Consider the following operations:

```
TO ADD.PROCEDURE :LIST
MAKE "TOTAL 0
REPEAT LENGTH :LIST
        [MAKE "TOTAL :TOTAL + FIRST :LIST
         MAKE "LIST BF :LIST]
END

TO ADD :LIST
IF EMPTYP :LIST [OP 0]
   [OP (FIRST :LIST) + ADD BF :LIST]
END
```

Each is designed to add a list of numbers. The first is a procedure (it doesn't return anything) that adds each element of its input to a global variable named **TOTAL**. In order for it to work, **:TOTAL** must be initialized to 0. During the course of the computation, the value of **:TOTAL** changes lots of times, depending on the number of elements of the list. After **REPEAT** has finished its job, the value of **:TOTAL** has the desired result stored away as its value. If we want to see it, we can then type in:

```
SHOW :TOTAL
```

ADD, on the other hand, simply computes the value, and returns the sum as a value. Of course, we have the option of saving the value returned by **ADD** by executing the statement

MAKE "TOTAL ADD *list of numbers*

If we want to use the result computed by **ADD.PROCEDURE**, we can reference the value of the global variable **:TOTAL**. If we want to use the value returned by **ADD**, we use the expression **ADD** *list* wherever it is needed. For example, to compute the average of a list of numbers in the functional style, we would use the following operation:

```
TO AVERAGE :LIST
OP (ADD :LIST) / (LENGTH :LIST)
END
```

A procedural version of **AVERAGE** might look like this:

```
TO AVERAGE.PROCEDURE :LIST
ADD.PROCEDURE :LIST
MAKE "AVERAGE :TOTAL / (LENGTH :LIST)
END
```

The functional version returns the average, whereas the procedural version leaves the value in a global variable named **AVERAGE**. Note that **AVERAGE.PROCEDURE** must know the name of the global variable in which the sum is saved.

In general, functional programming is much cleaner. Functions can be treated as black boxes, which return a certain result. Nobody cares how the value is computed. Procedural programming requires that we know what variable or variables were modified in order to use the result(s).

† **Exercise 6.8.1.** Explain the difference between **AVERAGE.PROCEDURE** and **ANOTHER.AVERAGE**, where **ANOTHER.AVERAGE** is defined as follows:

```
TO ANOTHER.AVERAGE :LIST
LOCAL "TOTAL
ADD.PROCEDURE :LIST
MAKE "AVERAGE :TOTAL / LENGTH :LIST
END
```

Exercise 6.8.2. Using the following definition, explain why the execution of **AVE.PROCEDURE [2 4 6]** fails.

```
TO AVE.PROCEDURE :LIST
LOCAL "TOTAL
MAKE "TOTAL 0
REPEAT LENGTH :LIST [ MAKE "TOTAL :TOTAL + FIRST :LIST
                     MAKE "LIST BF :LIST]
MAKE "AVERAGE :TOTAL / (LENGTH :LIST)
END
```

6.9 Sets as Lists

In this section we implement the basic operations of sets in Logo. Local variables are used heavily in defining these operations, and their mutually recursive cousins are examined for comparison.

A **set** is a collection of objects, such as numbers, words, or other sets. The objects in the collection are called the **elements** of the set. A set is denoted by writing the elements between braces, or curly brackets. For example, the set containing the letters *a*, *b*, and *c* is written as {a, b, c}.

A list also is a collection of objects, but there are some important differences between sets and lists:

1. The same object cannot appear twice in a set.
2. There is no order associated with the objects in a set.

These small differences make sets and lists very different kinds of objects. For example, **FIRST** is a meaningless operation as far as sets are concerned because the elements are not ordered. In addition, two lists are equal if their respective elements are equal. This says that the list **[A B]** is not equal to the list **[B A]**. However, two sets are equal if every element of each is an element of the other. That is, {a, b} is equal to {b, a}. The operations associated with sets are different than those associated with lists.

Set Primitives

IS.EMPTYSET *set*
IS.EMPTYSET is a recognizer that returns **TRUE** if its single input value is the empty set.

IS.ELEMENT.OF *object set*
IS.ELEMENT.OF is a recognizer that requires two input values. It returns **TRUE** if the first input is an element of the second input. The second input must be a set.

> **GET.ELEMENT** *set*
> **GET.ELEMENT** is a selector that returns an arbitrary element of a set.
>
> **DELETE.ELEMENT** *object set*
> **DELETE.ELEMENT** is a selector that requires a set and a set element as its inputs. It returns a set identical to its set input, but with the specified element deleted.
>
> **ADD.ELEMENT** *object set*
> **ADD.ELEMENT** is a constructor that requires a set element and a set as its only arguments. It returns a set constructed by adding the specified element to the specified set.
>
> **MKEMPTYSET**
> **MKEMPTYSET** is a constructor that takes no input values. It returns the empty set.

Logo does not provide sets as objects with which to compute, but we can represent sets as lists. For example, the set {a, b, c} can be represented by the list **[A B C]**. Then we can define the set operations just given in terms of Logo functions and procedures that operate on lists. The empty set is conveniently represented by the empty list.

Note that several different lists can represent the same set. That is, the set {a, b, c} could be represeted by the list **[A B C]** or by the list **[C A B]**. Also, note that there are lists that are not the representation of any set. That is, the list **[A B A]** does not represent any set because of the repetition of an element.

Using this list representation for sets, we can define the primitive operations in terms of list functions as follows:

    TO IS.EMPTYSET :LIST
    OP EMPTYP :LIST
    END

    TO IS.ELEMENT.OF :OBJECT :LIST
    OP MEMBERP :OBJECT :LIST
    END

```
TO GET.ELEMENT :LIST
OP NTH :LIST (1 + RANDOM LENGTH :LIST)
END

TO DELETE.ELEMENT :OBJ :LIST
OP ALL.BUT.FIRST :LIST :OBJ
END*

TO ADD.ELEMENT :OBJ :LIST
IF MEMBERP :OBJ :LIST [OP :LIST]
   [OP FPUT :OBJ :LIST]
END

TO MKEMPTYSET
OP [ ]
END
```

The assumption in **GET.ELEMENT** is that it returns a random element, which is in keeping with the idea that there is no order associated with the elements of a set. (Note that "getting" an element does not delete the element from the set).

Adding an element is done by putting it onto the front of the list representing the set if it is not already there. Although this does not adhere strictly to the principles of sets, it serves as a convenient way to add elements. In using the **ADD.ELEMENT** function with sets, you should ignore the fact that elements are added onto the front of the list.

MKEMPTYSET is used to return the empty set. We use this function in the construction of other sets. For example, we could define a primitive function named **MKSET**, which takes a list of objects and returns a list representing a set whose elements are those objects, as follows:

```
TO MKSET :LIST
IF EMPTYP :LIST [OP MKEMPTYSET]
   [OP ADD.ELEMENT FIRST :LIST MKSET BUTFIRST :LIST]
END
```

---

*ALL.BUT.FIRST is discussed in Exercise 5.4.3.

We can now define other operations on sets in terms of these primitive operations. When defining such operations, we should use the primitive operations described and forget that sets are represented by lists.

The notion of subset has no corresponding interpretation for lists.

> A set B is said to be a **subset** of a set C if every element of B is also an element of C.

The equality of two sets is defined in terms of the subset operation.

> Two sets B and C are said to be **equal** if B is a subset of C and C is a subset of B.

We can construct new sets from existing ones by forming the union or intersection of two sets. These operations are defined as follows.

> The **union** of two sets B and C is the set of all elements that are in either B or C. The union is denoted by $B \cup C$.
>
> The **intersection** of two sets B and C is the set of all elements that are in both B and C. The intersection is denoted by $B \cap C$.

For example,

$\{1, 2, 3, 4\} \cup \{3, 5, 6\}$ is the set $\{1, 2, 3, 4, 5, 6\}$

$\{1, 2, 3, 4\} \cap \{3, 5, 6\}$ is the set $\{3\}$

The following definition gives the descriptions of the set operations using the terminology of Logo.

> ## Additional Set Operations
>
> **IS.SUBSET** *set set*
> **IS.SUBSET** is a binary function that takes sets as its input values. It returns **TRUE** if the first input is a subset of the second input and **FALSE** otherwise.
>
> **EQUAL.SETS** *set set*
> **EQUAL.SETS** is a binary function that takes sets as its input values. It returns **TRUE** if the two sets are the same and **FALSE** otherwise.
>
> **UNION** *set set*
> **UNION** is a binary function that takes sets as its input values. It returns the union of the two sets.
>
> **INTERSECTION** *set set*
> **INTERSECTION** is a binary function that takes sets as its input values. It returns the intersection of the two sets.

IS.SUBSET must check to see if the first input is the empty set. If it is, then **IS.SUBSET** should return **TRUE**, since the definition is satisfied for this special case. If the first input is not empty, then **IS.SUBSET** must check to see if each element of the first set is an element of the second set.

The Logo definition of **IS.SUBSET** as well as the definitions of the other operations just described should use the primitive set operations described on page 213 and 214. We can describe the method of computation as follows:

    1. If the first set is empty, return **TRUE**.

    2. Otherwise, check to see if the elements of the first set are elements of the second set as follows:

        a. Get an element of the first set.

        b. If the element in (a) is in the second set, see if the remaining elements of the first set are elements of the second. Otherwise, return **FALSE**.

Here is the definition of **IS.SUBSET**.

```
TO IS.SUBSET :SET1 :SET2
LOCAL "ELEMENT
IF IS.EMPTYSET :SET1 [OP "TRUE]
MAKE "ELEMENT GET.ELEMENT :SET1
IF IS.ELEMENT.OF :ELEMENT :SET2
   [OP IS.SUBSET DELETE.ELEMENT :ELEMENT :SET1 :SET2]
   [OP "FALSE]
END
```

IS.SUBSET uses a local variable named **ELEMENT** to save temporarily the element selected from **:SET1**. This is necessary because the element is used by both **IS.ELEMENT.OF** and **DELETE.ELEMENT**, and **GET.ELEMENT** randomly picks an element out of the set input. We should not expect **GET.ELEMENT** to get the same element on two successive calls.

If the chosen element is also an element of **:SET2**, then **IS.SUBSET** checks to see if the remaining elements of **:SET1** form a subset of **:SET2**. That is, **IS.SUBSET** returns the result obtained by calling **IS.SUBSET** after deleting the element from **:SET1**.

The **IS.SUBSET** function can alternatively be defined by two mutually recursive functions:

```
TO IS.SUBSET :SET1 :SET2
IF IS.EMPTYSET :SET1 [OP "TRUE]
   [OP CHECK.ELEMENTS.OF (GET.ELEMENT :SET1) :SET1
   :SET2]
END
```

The recognizer, **CHECK.ELEMENTS.OF**, performs the task specified in part (b) of the algorithm. It has three inputs. The first input value is an element of the first set and the other inputs are the original input sets to **IS.SUBSET**.

```
TO CHECK.ELEMENTS.OF :ELEMENT :SET1 :SET2
IF IS.ELEMENT.OF :ELEMENT :SET2
   [OP IS.SUBSET (DELETE.ELEMENT :ELEMENT :SET1) :SET2]
   [OP "FALSE]
END
```

The algorithm for the **UNION** operation can be described using a similar algorithm:

1. If the first set is empty, return the second set.

2. Otherwise, construct a set as follows:

    a. Get an element of the first set.

    b. Add the element found in (a) to the union of the first set (with the element deleted) and the second set.

Thus, the definition of **UNION** can be given as follows:

```
TO  UNION :SET1 :SET2
LOCAL "ELEMENT
IF IS.EMPTYSET  :SET1 [OP :SET2]
MAKE "ELEMENT GET.ELEMENT :SET1
OP ADD.ELEMENT :ELEMENT
                UNION (DELETE.ELEMENT :ELEMENT :SET1)
                      :SET2
END
```

† **Exercise 6.9.1.** Define the **EQUAL.SETS** function described on page 217.

Exercise 6.9.2. Define the **INTERSECTION** function described on page 217 using a local variable.

† **Exercise 6.9.3.** Define the **INTERSECTION** function described on page 217 using a pair of mutually recursive functions.

Exercise 6.9.4. Define a Logo function named **SETDIFF**, which represents the following set operation:

> **SETDIFF** is a binary function whose inputs must be sets. It returns a set whose elements are all the elements of the first set that are not elements of the second set. For example,

SETDIFF [1 2 3 5 6 7] [1 3 5 8] returns **[2 6 7]**

† **Exercise 6.9.5.** Define a primitive function named **IS.SET** that takes a single input value. It returns true if its input value is a legitimate representation of a set. This function will know about the underlying list representation for sets. It must make sure that its input is a list and that there are no duplicate entries. For example,

IS.SET "X	returns	FALSE
IS.SET [X Y Z]	returns	TRUE
IS.SET [X Y X Z]	returns	FALSE
IS.SET []	returns	TRUE

Chapter 7
Programming With Graphical Objects

7.1 Introduction

The graphics capability of Logo allows us to draw objects such as squares and triangles, but it does not allow us to manipulate them as objects. That is, we can draw the picture of a square but there is no command to move it to a different location. We can erase it from the screen and draw a new picture, but this is a painstaking procedure. The Logo turtle is a graphical object. It can be moved, rotated, and made invisible by executing simple statements.

This chapter examines ways to create environments in which to manipulate graphical objects freely. The key is to view the object as a collection of properties rather than as a picture that appears on the screen. The ideas are made concrete by building a world of multiple turtles and developing a graphical solution to the tower of Hanoi problem. Developing environments for programming with geometric objects such as squares and triangles are left as projects. You should attempt at least one to develop your understanding fully.

7.2 The Turtle as a Collection of Properties

The turtle has the following properties: position, heading, penstate, and visibility state. We communicate with the turtle by executing a procedure that modifies the value of one or more of its properties (**SETPOS**, **FD**, **SETPC**) or by executing a function that returns the value of one of its properties (**XCOR**, **SHOWNP**, **HEADING**). Whenever a property is modified, the screen is updated to reflect the change in the turtle's properties. The idea is to separate the concept of the turtle from the actual graphical display. Thus, we define a turtle as follows.

> **A turtle** is a collection of properties.

We can represent these property values as a list whose first element is a list representing its position, whose second element is a number representing its current heading, whose third element is a list representing its pen state, and whose fourth component is a truth value representing its visibility state.

That is, a turtle can be completely described by a list of the following form:

[*position heading penstate visibility state*]

The list

[[0 0] 45 [PENDOWN RED] TRUE]

can be interpreted as a visible turtle whose position is the point associated with the ordered pair (0, 0), whose heading is 45°, and which has a pen whose color is red and is ready for drawing.

The list

[[100 50] 180 [PENUP BLUE] FALSE]

represents an invisible turtle at the point (100, 50) with a heading of 180° holding a pen with blue ink in the up position.

We can envision this list being tucked away inside the computer. The procedures that modify a property of a turtle can be thought of as programs that replace this list with a new list reflecting the modification. The functions that return a particular property simply return the appropriate item of the list.

To make this idea more concrete, we assume the existence of a primitive function named **GET.TURTLE**, which simply returns the list of properties of the turtle, and a procedure named **SET.TURTLE**, which takes a list representing a turtle and installs it as the turtle. **SET.TURTLE** is responsible for updating the graphics screen as well as installing the list representing the turtle.

Given this underlying representation, we can define the selector functions in terms of list functions. For example, **POS**, **HEADING**, **PEN**, and **SHOWNP** can be defined as

```
TO POS
OP ITEM 1 GET.TURTLE
END

TO HEADING
OP ITEM 2 GET.TURTLE
END

TO PEN
OP ITEM 3 GET.TURTLE
END

TO SHOWNP
OP ITEM 4 GET.TURTLE
END
```

ITEM is a Logo primitive function that takes a positive integer and a list as inputs. It returns the specified element of the list.

In order to isolate the construction of a turtle in a single place, we define a function named **MK.TURTLE** that takes the four property values as its inputs and returns a list of these values:

```
TO MK.TURTLE :LOC :DIR :PENSTATE :VISIBILITY
OP (LIST :LOC :DIR :PENSTATE :VISIBILITY)
END
```

The definitions of the procedures that modify a property of the turtle create a turtle with the desired properties and install it using **SET.TURTLE**. For example, **SETPOS** could be defined as follows:

```
TO SETPOS :LOC
SET.TURTLE MK.TURTLE :LOC HEADING PEN SHOWNP
END
```

Exercise 7.2.1. Write the list representations for each of the following turtles.

a. A visible turtle that is located at the point associated with the ordered number pair (-30, 100), facing southeast, with a pen whose ink color is violet and ready for drawing.

b. A visible turtle located at (100, -30), facing west, with an orange pen that does not leave a trail when it moves.

† **Exercise 7.2.2.** Define the procedures **SETHEADING, SETPC, SHOWTURTLE, HIDETURTLE,** and **PENDOWN** using **SET.TURTLE**.

† **Exercise 7.2.3.** Define the functions **PENCOLOR, XCOR,** and **YCOR** using **GET.TURTLE**.

7.3 Multiple Turtles

This section describes how to create a multiple-turtle environment. The idea is to enable the user to create as many turtles as he or she desires and to direct requests to particular turtles. The requests will be the familiar ones: **FD, RT, SETPOS,** and so on.

The user will be provided with a procedure to create turtles and a procedure that directs requests to a particular turtle. All turtles will be named so that the particular one requested to perform an action can easily be specified.

The procedure used to create a turtle is named **HATCH**. It requires a word denoting the name of the turtle as input.

The procedure that directs requests to a particular turtle is named **SENDTO**. **SENDTO** has the name of a turtle as its input. A **SENDTO** statement is executed prior to the requests directed to the named turtle. For example, to create turtles named **ADAM** and **EVE**, you enter the following:

```
?HATCH "ADAM
?HATCH "EVE
```

If you want **ADAM** to draw the picture of a square with one corner at the point **[-50 0]**, you would continue with

```
?SENDTO "ADAM
?PENUP
?SETPOS [-50 0]
?PENDOWN
?SQUARE 25
```

This tells the turtle named **ADAM** to move itself to the desired point with its pen in the "up" position, lower the pen, and execute the **SQUARE** procedure. You can then switch your attention to **EVE** by executing

```
?SENDTO "EVE
?PENUP
?SETPOS [50 0]
?PENDOWN
?REPEAT 3 [FD 30 RT 120]
?FD 30
?RECTANGLE 20 40
```

This sequence of statements produces the picture in Figure 7.1 on the screen.

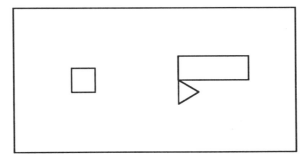

Figure 7.1

226 *Programming With Graphical Objects*

In addition to the **HATCH** and **SENDTO** procedures, we provide a procedure named **BANISH**, which eliminates turtles, and a function named **TURTLES**, which returns a list of names of currently existing turtles:

```
?SHOW TURTLES
[ADAM EVE]
?BANISH "ADAM
? SHOW TURTLES
[EVE]
```

The three procedures and one function are defined next.

User-Level Procedures and Function

Procedures

HATCH *name*
HATCH is a procedure that creates a turtle and gives it the name *name* . A turtle name can be any Logo word except words used to denote numbers.

SENDTO *name*
SENDTO makes the turtle with indicated name respond to all subsequent requests.

BANISH *name*
BANISH removes the named turtle from the environment.

Function

TURTLES
TURTLES is a function with no inputs. It returns a list of currently existing turtles.

Hatching a turtle involves creating a list of properties, associating its name with the list, and adding the name to the list of currently existing turtles.

We first need to decide what the initial properties of a newly hatched turtle should be. Here are two reasonable choices:

1. The new turtle will have the default properties of the real Logo turtle.

2. The new turtle will inherit the current properties of the real Logo turtle.

Arbitrarily choosing the second alternative, the list denoting the turtle is computed by the following expression:

(LIST POS HEADING PEN SHOWNP)

We assume the existence of procedures named **SET.TURTLE** and **ADD.TURTLE.NAME** to do the remaining jobs and arrive at the following definition for **HATCH**:

```
TO HATCH :NAME
SET.TURTLE :NAME MK.TURTLE POS HEADING PEN SHOWNP
ADD.TURTLE.NAME :NAME
END
```

The definition of **MK.TURTLE** is the one given in the previous section. The **SET.TURTLE** procedure is responsible for saving the newly created turtle by associating its name with the list of properties. The **ADD.TURTLE.NAME** procedure saves the new name with the names of the other turtles. The definition of **ADD.TURTLE.NAME** is given at the end of this section, and the definition of **SET.TURTLE** is given in the next section.

Auxillary Procedures Needed by HATCH

SET.TURTLE *name turtle*
SET.TURTLE is a procedure that associates the indicated name with the list representing the turtle.

ADD.TURTLE.NAME *name*
ADD.TURTLE.NAME is a procedure that saves the name in the list containing all turtle names.

The real turtle simulates the responder. To do this, the **SENDTO** procedure has the real Logo turtle take on the properties of the responder. For example, if we enter

    SENDTO "ADAM

the property values associated with the turtle named **ADAM** are given to the real turtle. As long as **ADAM** is the responder, all requests affect the properties of the real Logo turtle and not the list of properties of **ADAM**. The list associated with **ADAM** is not updated until there is a change of responder.

Since the list of properties associated with the responder is not updated after every request, **SENDTO** must first update the property list of the current responder before the real turtle takes on the properties of a new responder.

The first job is performed by a procedure named **UPDATE.CURRENT.RESPONDER** and the second, by a procedure named **SIMULATE.NEW.RESPONDER**. The definition of **SENDTO** can be given by:

    TO SENDTO :NAME
    UPDATE.CURRENT.RESPONDER
    SIMULATE.NEW.RESPONDER :NAME
    END

Updating the properties of the current responder is performed by first checking to see if a current responder exists (there will not be a responder the first time **SENDTO** is executed). If there is one, we construct a list representing the properties of the current responder (the **MK.TURTLE** expression), get the name of the current responder (**GET.RESPONDER.NAME**), and associate the list with the name (**set.turtle**). If there is no current responder, nothing is done. Thus, we have the following definition of **UPDATE.CURRENT.RESPONDER**:

    TO UPDATE.CURRENT.RESPONDER
    IF EXISTS.RESPONDERP
      [SET.TURTLE GET.RESPONDER.NAME
              MK.TURTLE POS HEADING PEN SHOWNP]
    END

Giving the real turtle the properties of the new responder involves setting the position, heading, penstate, and visibility state of the real Logo turtle to the corresponding property values of the new responder. We first put the pen up so that no lines will be drawn:

```
TO SIMULATE.NEW.RESPONDER :NAME
PU
SETPOS GET.POS :NAME
SETH GET.HEADING :NAME
SETPEN GET.PENSTATE :NAME
IF GET.VISIBILITY :NAME [ST] [HT]
MAKE.RESPONDER :NAME
END
```

The properties of the new responder are obtained by **GET.POS**, **GET.HEADING**, **GET.PENSTATE**, and **GET.VISIBILITY**. The last statement calls a procedure named **MAKE.RESPONDER** to store away the name of the new responder.

To complete the definition of **SENDTO** we need to define **GET.POS**, **GET.HEADING**, **GET.PENSTATE**, **GET.VISIBILITY**, **EXISTS.RESPONDERP**, **GET.RESPONDER.NAME**, and **MAKE.RESPONDER**. The first four depend on how we choose to associate names of turtles with their lists of properties. The other three depend on how we decide to keep track of which turtle is the current responder.

Auxillary Procedures Needed by SENDTO

GET.POS *name*
GET.POS is a function that returns the position of the named turtle.

GET.HEADING *name*
GET.HEADING is a function that returns the heading of the named turtle.

GET.PENSTATE *name*
GET.PENSTATE is a function that returns the penstate of the named turtle.

> **GET.VISIBILITY** *name*
> **GET.VISIBILITY** is a function that returns the visibility state of the named turtle.
>
> **EXISTS.RESPONDERP**
> **EXISTS.RESPONDERP** is a function that returns **TRUE** if a current responder exists and **FALSE** otherwise.
>
> **GET.RESPONDER.NAME**
> **GET.RESPONDER.NAME** is a function that returns the name of the current responder.
>
> **MAKE.RESPONDER** *name*
> **MAKE.RESPONDER** is a procedure that remembers the name of the new responder.

The **BANISH** procedure must do two things also. It must eliminate the name of the turtle from the list of existing turtles, and it must delete the list representing the turtle as well as the association of the name of the turtle and the list.

```
TO BANISH :NAME
DELETE.TURTLE.NAME :NAME
DELETE.TURTLE :NAME
END
```

DELETE.TURTLE and **DELETE.TURTLE.NAME** are described next.

> **Auxillary Procedures Needed by BANISH**
>
> **DELETE.TURTLE** *name*
> **DELETE.TURTLE** is a procedure that deletes the named turtle.
>
> **DELETE.TURTLE.NAME** *name*
> **DELETE.TURTLE.NAME** is a procedure that removes the name from the list of all turtles.

There are a number of procedures and functions that need to be defined. Their definitions depend upon how we save the name of the current responder, how we associate the name of a turtle with its list of properties, and how we save the list of existing turtles.

The procedures **SET.TURTLE** and **DELETE.TURTLE** as well as the functions **GET.POS, GET.HEADING, GET.PENSTATE,** and **GET.VISIBILITY** depend on how we associate the name of a turtle with its list of properties. The function **TURTLES** and the procedures **DELETE.TURTLE.NAME** and **ADD.TURTLE.NAME** depend on how the list of existing turtles is saved. Finally, the functions **EXISTS.RESPONDERP** and **GET.RESPONDER.NAME** and the procedure **MAKE.RESPONDER** depend on how we save the name of the turtle dubbed as the current responder.

There are three ways we can save the desired information: use global variables, use a symbol table, or use property lists. We use global variables to store the name of the current responder and the list of currently existing turtles. Property lists, described in the next section, are more appropriate for representing the properties of turtles. We define the functions and procedures dealing with the current responder and the list of existing turtles in this section and give the definitions of the remaining functions and procedures in the next section.

If we name the global variable representing the list of turtle names, **NAMESOFEXISTINGTURTLES**, the definitions of **TURTLES, ADD.TURTLE.NAME,** and **DELETE.TURTLE.NAME** can be given by

```
TO TURTLES
IF NAMEP "NAMESOFEXISTINGTURTLES
  [OP :NAMESOFEXISTINGTURTLES]
  [OP []]
END

TO ADD.TURTLE.NAME :NAME
IF NAMEP "NAMESOFEXISTINGTURTLES
  [MAKE "NAMESOFEXISTINGTURTLES
        FPUT :NAME :NAMESOFEXISTINGTURTLES]
  [MAKE "NAMESOFEXISTINGTURTLES LIST :NAME]
END
```

```
TO DELETE.TURTLE.NAME :NAME
IF NAMEP "NAMESOFEXISTINGTURTLES
   [MAKE "NAMESOFEXISTINGTURTLES
          ALLBUT :NAME :NAMESOFEXISTINGTURTLES]
END
```

We choose the name of the global variable to contain a large number of characters in hopes that the user will never use it in his or her programs. Funny things might happen if it is used.

In the beginning the variable **:NAMESOFEXISTINGTURTLES** does not exist, so we need to check for its existence. It is created the first time **ADD.TURTLE.NAME** is executed. **TURTLES** simply returns an empty list if the global variable does not exist, because there are no turtles. **DELETE.TURTLE.NAME** simply ignores the request to delete a turtle if no turtles exist. (See Exercise 5.4.5 for the definition of **ALLBUT**.)

Exercise 7.3.1. Using a global variable named **CURRENTRESPONDINGTURTLE** to store the name of the current responder, give the definitions of the following procedures and functions.

† a. **EXISTS.RESPONDERP**
   b. **GET.RESPONDER.NAME**
   c. **MAKE.RESPONDER**

7.4 Property Lists

Property lists are a feature of Logo that allows the global representation of information without most of the drawbacks of global variables. In the multiple-turtle program, we could use the word **ADAM** as the name of a global variable and save the list of **ADAM**'s properties as its value. We would be living dangerously doing this because the user will be choosing the names of turtles that will be turned into global variables and we should not assume that he or she understands the problems with global variables.

The idea encapsulated by property lists is that a single object may have a number of properties, and the value associated with a particular property will change over a period of time. For example, a turtle has position, heading, penstate, and visibility properties. When we **HATCH** a turtle named **ADAM** in the multiple-turtle environment we want the word **ADAM** to take on these additional properties. We can do this by using the Logo primitive **PPROP**.

PPROP (short for "put property" and value) allows us to associate a property name and value to any word that is not a number. For example, to associate the list **[0 50]** with the **POSITION** property of the word **ADAM**, we enter

> ?PPROP "ADAM "POSITION [0 50]

Changing the value of a property is also performed by executing a **PPROP** statement. We can obtain the value of a particular property of a particular word using the **GPROP** (short for "get property" value) function:

> ?SHOW GPROP "ADAM "POSITION
> [0 50]

The number and kinds of properties may change over time as well. For example, a person always has an age, height, and weight, whereas other properties, like a job, exist only for a portion of one's life. A property and its value can be removed by using the **REMPROP** procedure.

The collection of property names and values that are currently associated with a word is called the word's **property list**. The entire property list (a list of name value pairs) of a particular word is returned by the **PLIST** primitive.

Property List Primitives

Procedures

PPROP *word propertyname value*
PPROP has three inputs. It associates the *value* input with the *propertyname* and puts the association on the property list of *word*.

REMPROP *word propertyname*
REMPROP removes the named property from the property list of *word*.

> **Functions**
>
> **GPROP** *word propertyname*
> **GPROP** returns the value associated with *property name* from the property list of *word*. If there is no such property, the empty list is returned.
>
> **PLIST** *word*
> **PLIST** returns the property list of *word*. The property list is a list of pairs. Each pair consists of a *propertyname* and the value currently associated with that property. Only user-defined properties are returned.

The property lists of words are maintained at the global level. Properties of a word cannot be made local to a procedure. This eliminates the problem encountered with global variables. That is, the use of a local variable by the same name as a global variable may serve to block access to the value of the global variable. Using property lists eliminates naming conflicts and problems that can occur with global variables because they are global entities only.

Instead of using global variables to hold onto the list of property values, we create four properties for the word naming the turtle: a position property, a heading property, a penstate property, and a visibility property. For example, we can define **SET.TURTLE** by

```
TO SET.TURTLE :NAME :LIST
PPROP :NAME "POSITION ITEM 1 :LIST
PPROP :NAME "HEADING ITEM 2 :LIST
PPROP :NAME "PENSTATE ITEM 3 :LIST
PPROP :NAME "VISIBILITY ITEM 4 :LIST
END
```

With this representation of a turtle, we can define **GET.POS**, **GET.HEADING**, **GET.PENSTATE**, and **GET.VISIBILITY** as follows:

```
TO GET.POS :TURNAME
OP GPROP :TURNAME "POSITION
END
```

```
TO GET.HEADING :TURNAME
OP GPROP :TURNAME "HEADING
END

TO GET.PENSTATE :TURNAME
OP GPROP :TURNAME "PENSTATE
END

TO GET.VISIBILITY :TURNAME
OP GPROP :TURNAME "VISIBILITY
END
```

Exercise 7.4.1. Define the **DELETE.TURTLE** procedure described on page 230.

Exercise 7.4.2. Eliminate all global variables in the multiple-turtle environment by following each set of directions.

† a. Redefine **DELETE.TURTLE.NAME, ADD.TURTLE.NAME,** and **TURTLES** so that property lists are used rather than the global variable **:CURRENTRESPONDINGTURTLE**.

b. Redefine **EXISTS.RESPONDERP, GET.RESPONDER.NAME,** and **MAKE.RESPONDER** so that property lists are used rather than the global variable **:NAMESOFEXISTINGTURTLES**.

Exercise 7.4.3. Define a procedure named **TURTLE.WORLD** that displays the properties of all existing turtles. The properties of each turtle should appear on a separate line with the first line, giving the properties of the current responder.

† **Exercise 7.4.4.** Modify the definition of **HATCH** so that whenever a new turtle is created it becomes the responder.

7.5 Programming with Multiple Turtles

Writing programs with multiple turtles is no different than writing programs for a single turtle, although you can perform many programming tasks more easily using multiple turtles. For example, having two distinct squares drawn (almost) simultaneously is awkward with one turtle but very straightforward using multiple turtles.

Assuming that we have **HATCH**ed turtles named **ADAM** and **EVE**, we will have **ADAM** draw the first side of one square, have **EVE** draw the first side of another square, then have **ADAM** do another piece, and so on. If it happens fast enough, it will appear that two squares are being drawn simultaneously. We can define a procedure to do this as follows:

```
TO TWO.SQUARE :SIZE
REPEAT 4 [SENDTO "ADAM FD :SIZE RT 90
          SENDTO "EVE  FD :SIZE RT 90]
END
```

We could define **TWO.SQUARE** to have additional inputs as well. For example, we could require two size values if we wanted the squares to have different side lengths. Or we could add inputs that gave the locations for the squares as well.

Rather than limiting the action to these particular turtles, we can make the names of the turtles doing the action input values:

```
TO TWO.TURTLE.SQUARE :TUR1 :TUR2 :SIZE
REPEAT 4 [SENDTO :TUR1 FD :SIZE RT 90
          SENDTO :TUR2 FD :SIZE RT 90]
END
```

TWO.TURTLE.SQUARE requires the names of two turtles as well as the size of the square as inputs. Now the named turtles do the square drawing.

The multiple-turtle environment does not allow us to view each turtle as an independent object. We should not have to wait for the completion of a request to one turtle before issuing a request to another turtle. For example, if we want **ADAM** and **EVE** to draw a square and a triangle at the same time, we should be able to say

"Hey, Adam, draw me a square."
"Hey, Eve, draw me a triangle."

in quick succession. **EVE** may get the triangle request slightly after **ADAM** gets the square request, but if they are truly independent **EVE** should not have to wait until **ADAM** is done before embarking on the triangle project.

This capability is referred to as **multiprocessing**. In a multiprocessing environment you do not need to worry about breaking up a task in small pieces in order to get the impression that independent tasks by independent objects are happening at the same time.

† **Exercise 7.5.1.** Define a procedure named **TURTLE.SQUARES** that takes two turtle names, two locations, and two numbers as inputs. **TURTLE.SQUARES** has each of the named turtles draw a square. The turtle named first draws a square starting at the first location input with a size given by the first number input. The turtle named second draws a square starting at the second location input with a size given by the second number input. The squares should be drawn simultaneously.

Exercise 7.5.2. Define a procedure named **SQUARE.TRIANGLE** that takes two turtle names and a number as inputs. It has the first turtle draw a square and the second turtle draw a triangle. The number input is used as the length of a side of each figure. The drawing should appear as if both figures are being drawn simultaneously.

† **Exercise 7.5.3.** Define a procedure named **ALL.SQUARE** that takes a list of turtle names and a number as inputs. Each named turtle will draw a square (the length of each side is given by the number) in the order their names appear in the list. That is, the squares will not be drawn simultaneously.

Exercise 7.5.4. Define a procedure named **SIMUL.SQUARES** that takes a list of turtle names and a number as inputs. Each named turtle draws a square. The squares should appear to be drawn simultaneously. That is, the first turtle in the list draws a side and then waits until all the other turtles draw a side before the second side is drawn.

† **Exercise 7.5.5.** *The Four Bugs Problem.* Define a program that takes four turtle names as inputs. The program simulates the four turtles chasing each other by repeating the following actions forever:

1. The first turtle takes a small step (try 5) toward the second turtle.
2. The second turtle takes a small step toward the third turtle.
3. The third turtle takes a small step toward the fourth turtle.
4. The fourth turtle takes a small step toward the first turtle.

Exercise 7.5.6. Define a procedure named **BANISH.ALL** that banishes all existing turtles.

Exercise 7.5.7. Define a procedure named **DISPLAY.ALL.TURTLES** that displays each existing turtle by flashing each turtle's picture in turn for 2 seconds. (HINT: Use the **WAIT** primitive.)

Exercise 7.5.8. Define a procedure named **SHOW.ALL.TURTLES** that displays each existing turtle in turn for 2 seconds. It should continue to do this until a key is pressed on the keyboard.

Exercise 7.5.9. Write a program that simulates a turtle race. The race begins with up to 10 turtles lined up on the left edge of the screen and ends when one of the participants reaches the right edge of the screen. The program has a number, *n*, denoting the number of turtles in the race, as its only input. It creates *n* turtles evenly spaced at the left edge of the screen, moves each of them a random number of units (from 0 to 10) to the right, and announces the winner when one reaches the right side of the screen.

7.6 Projects: Geometric Figures as Graphical Objects

This section outlines projects that deal with geometric objects. The first one deals with squares.

† **Project 7.1: Square World**

Squares are graphical objects that we might want to move around the screen, rotate about a point, and grow or shrink (change the length of their sides). Each square has a name so that we can single out a particular square to carry out a particular request.

There are two areas of programming to think about. One area, the square-management area, consists of the procedures and functions that create new squares, delete squares, designate the responder, and inquire about the names of squares. The second area, the functionality area, consists of the procedures and functions to which we want squares to react: **GROW, ROTATE, SETLOC**, and so on.

In the square-management area, we need to define the following functions and procedures:

1. **CREATE.SQUARE** is a procedure with a name, a color, and a size as inputs. It creates a square with one corner at the point **[0 0]**.

2. **DELETE.SQUARE** is a procedure with a name as its input. It deletes the named square.

3. **SENDTO.SQUARE** is a procedure used to designate the current responder. It has the name of a square as its input.

4. **SQUARENAMES** is a function with no inputs. It returns a list of names of existing squares.

The user creates squares by entering expressions like the following:

```
?CREATE.SQUARE "BIG 3 50
?CREATE.SQUARE "TIMESQUARE 0 100
?CREATE.SQUARE "SMALL 2 25
```

He or she has squares perform requests by first singling out the current responder and then issuing the requests. For example, the requests

```
?SENDTO.SQUARE "TIMESQUARE
?ROTATE 45
?FWD 20
?SENDTO.SQUARE "SMALL
?GROW 20
```

tell **TIMESQUARE** to rotate itself 45° and **SMALL** to grow so that it has sides of 20 units each.

Here are the requests to which we would like squares to react:

R1. **FWD** is a procedure that causes a square to move itself forward. It has one input: the distance to move.

R2. **ROTATE** is a procedure that causes a square to rotate itself to the right. It has one input: a rotation angle.

R3. **SETLOC** is a procedure that causes a square to change its location. It has one input: a point (a list of two numbers representing the *x*- and *y*-coordinates of the new location).

R4. **GROW** is a procedure that causes a square to change its size. It has one input: a number indicating the new length of each side.

R5. **LOC** is a function that returns the location of the current responder. It has no inputs.

R6. **HDG** is a function that returns the orientation of the current responder. It has no inputs.

R7. **SIZE** is a function that returns the size of the current responder. It has no inputs.

R8. **COLOR** is a function that returns the color of the current responder. It has no inputs.

Using these requests as our guide, we see that a square must have the following properties:

Location: A point representing one corner of the square.

Orientation: An angle representing the heading of the square.

Size: The length of each side of the square.

Color: A number indicating the color of its sides.

Creating squares is performed by creating a property list on the word that names the square, displaying a picture of the square on the screen, and adding its name to a list of existing squares. The property list uses the inputs provided by the user for the name, size, and color properties, [0 0] as its location, and 0 as the default value for its heading.

Directing requests to a particular square is done by having a current responder, as we did for multiple turtles. Each request is implemented as a function or a procedure. If the request is a request to return the current value of one of its properties (such as size), the function returns the value associated with the current responder. If the request is one that requires the modification of the properties, the procedure does so and redraws the square to reflect the change on the display screen.

Note that the properties of squares are modified by the procedures implementing requests to perform some action. The turtle is not used to simulate a square, since a turtle and a square are different kinds of objects.

Squares are always visible, so we also need to display their pictures on the screen. Thus, **CREATE.SQUARE** can be given by:

```
TO CREATE.SQUARE :NAME :SIZE :COLOR
SET.SQUARE :NAME MK.SQUARE [0 0] 0 :SIZE :COLOR
ADD.TO.SQUARENAMES :NAME
END
```

MK.SQUARE creates the list of properties and displays the square:

```
TO MK.SQUARE :LOC :HDG :SIZE :COLOR
DRAW.SQUARE :LOC :HDG :SIZE :COLOR
OP (LIST :LOC :HDG :SIZE :COLOR)
END
```

SENDTO.SQUARE just takes the name of a square as its argument and updates the value of the global variable **:RESPONDER**.

Complete the project by doing the following:

    a. Define **SET.SQUARE**.
    b. Define **ADD.TO.SQUARENAMES**.
    c. Define **DRAW.SQUARE**.

d. Define a procedure named **ERASE.SQUARE,** which takes the location, heading, color, and size of a square as inputs and erases the picture of the square from the screen.
e. Define **SENDTO.SQUARE.**
f. Define procedures to handle the requests R1 through R8.

Project 7.2: Polygon World

Do Project 7.1, but allow the objects to be polygons with any number of sides. That is, the world may be populated with squares, triangles, pentagons, octagons, and so on. (*Hint*: Add another property, namely, the number of sides of the polygon object.)

7.7 The Turtle as a Procedure

In an attempt to streamline our interaction with multiple turtles, we would like to enter expressions such as

    ?ADAM SQUARE 30   (1)
    ?EVE TRIANGLE 20   (2)

rather than

    ?SENDTO "ADAM SQUARE 30   (3)
    ?SENDTO "EVE TRIANGLE 20   (4)

If we tried entering expressions (1) and (2) in Logo, we would get the following error messages:

    I DON'T KNOW HOW TO ADAM.
    I DON'T KNOW HOW TO EVE.

We obtain these errors since **ADAM** and **EVE** are not preceded by quotation marks, and hence they are viewed by Logo as the names of procedures. But this fact also leads us to the solution of how to allow statements like forms (1) and (2). All we need to do is to define procedures named **ADAM** and **EVE**, which are equivalent to **SENDTO** "**ADAM** and **SENDTO** "**EVE**, respectively. That is, we just add the following definitions to what we already have:

```
TO ADAM
SENDTO "ADAM
END

TO EVE
SENDTO "EVE
END
```

The time to add these procedures is when a new turtle is **HATCH**ed. We do not want to have to do this by hand every time we create a new turtle, so we modify the definition of **HATCH** to include defining them. The new version of **HATCH** is given by

```
TO HATCH :NAME
SET.TURTLE :NAME MK.TURTLE POS HEADING PEN SHOWNP
ADD.TURTLE.NAME :NAME
DEFINE :NAME LIST [ ] LIST "SENDTO WORD "" :NAME
END
```

The name of the turtle being created is used as the name of the procedure. The second input to **DEFINE** is the list

[[ ] [SENDTO *name*]]

where *name* is the name of the turtle. The empty list indicates that this new procedure will have no inputs.

The interaction with **ADAM** and **EVE** described earlier can now be performed as follows:

```
?HATCH "ADAM
?HATCH "EVE
?ADAM SETPOS [-100 0]
?EVE SETPOS [50 0] SQUARE 50
?ADAM SQUARE 50
```

Although this variation of directing requests to individual turtles uses the **SENDTO** procedure as before, it suggests that a turtle can be represented as a procedure. The inputs to the procedure can be viewed as the requests we want the turtle to perform. To represent a turtle by a procedure, the procedure itself must keep track of the turtle's properties. That is, the procedure itself should have a state (the properties of the turtle along with the current values of the properties). Whereas this is not possible in Logo, other languages such as Scheme and Common LISP allow such procedures.

Exercise 7.7.1. Modify your solution to Project 7.1 so that procedures are used to direct requests to a particular object.

Exercise 7.7.2. Modify your solution to Project 7.2 so that procedures are used to direct requests to a particular object.

7.8 Towers of Hanoi

The Towers of Hanoi problem consists of three pegs named A, B, and C. Peg A contains a stack of n different-sized disks arranged so that each disk rests on top of a larger disk (see Figure 7.2).

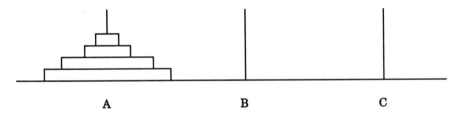

Figure 7.2 Towers of Hanoi

The problem is to move the stack of disks from peg A to peg C using the following rules:

1. Each move consists of moving the top disk from one peg to another peg.

2. A disk cannot be placed on top of a smaller one.

The first problem to confront before writing a program that accomplishes the task is to decide if the problem can be solved. In order to keep track of what disks we are moving we give them names: disk 1, disk 2, disk 3, ..., disk n. We also assume that disk 1 is the smallest, disk 2 the next smallest, and so on.

The solution to the problem is easy if the stack consists of one or two disks:

1. If $n = 1$, simply move disk 1 from peg A to peg C.

2. If $n = 2$:

   a. Move disk 1 from peg A to peg B.
   b. Move disk 2 from peg A to peg C.
   c. Move disk 1 from peg B to peg C.

With a little more effort we can figure out how to move a stack of three disks. This reasoning convinces us that the problem is solvable for some stacks, but it does not tell us if it is solvable for stacks of 17 or 153 disks. When the stack is four high the problem is more complicated, but rather than figuring out a sequence of moves that solves it for four we can prove that it is solvable for a four-high stack by reasoning as follows:

1. Since the problem is solvable for a three-disk stack, we know that we can move the top three disks on peg A onto peg B. That is, regard peg B as the finish peg for a three-high stack using peg C for temporary placement.

2. Next move disk 4 from peg A to peg C.

3. Since the problem is solvable for a three-disk stack we know that we can move the three disks on peg B to peg C using peg A for temporary placement.

We can now use the same reasoning to prove that the problem is solvable for a stack of five disks and then for six. You should soon be convinced that the problem is solvable for any number of disks.

Now let's try our hand at writing a program that produces a sequence of moves that solves the problem for an arbitrary number of disks. To represent the moving of a disk from one peg to another, we need to specify the name of the start peg and the name of the destination peg. Since we are permitted to move only the top disk, we do not need to include the name of the disk. For example, "move the disk on top of peg B to peg A" can be represented as the list

[B A]

The sequence of such moves to solve the problem can be represented as a list of moves. Thus, the solution to the three-disk problem can be represented as:

[[A C] [A B] [C B]   (1)
 [A C]   (2)
 [B A] [B C] [A C]]   (3)

The program follows the logic formulated earlier. That is, given the names of the start peg, the finish peg, and the temporary-placement peg, we solve the n-disk problem as follows:

1. Solve the n - 1 disk problem using the start peg as the start, the temporary-placement peg as the finish, and the finish peg for temporary-placement peg.

2. Move the top disk on the start peg to the finish peg.

3. Solve the n - 1 disk problem using the temporary-placement peg as the start, the finish peg as the finish, and the start peg for temporary-placement.

The function that returns the list of moves is named **SOLVE.HANOI**. It will have four inputs: the name of the start peg, the name of the finish peg, the name of the temporary-placement peg, and the number of disks for the particular problem we want to solve.

If there is only one disk involved, **SOLVE.HANOI** returns a list containing the single move. Otherwise, **SOLVE.HANOI** constructs a list by merging the list of moves needed to solve the n - 1 problem described in (1), the single-move list needed to solve the single-disk problem described by (2), and the list needed to solve the n - 1 problem described in (3). Here it is:

```
TO SOLVE.HANOI :STARTPEG :FINISHPEG :TEMPPEG :N
IF :N=1 [OP MK.MOVE :STARTPEG :FINISHPEG]
OP (SE SOLVE.HANOI :STARTPEG :TEMPPEG :FINISHPEG :N-1
       MK.MOVE :STARTPEG :FINISHPEG
       SOLVE.HANOI :TEMPPEG :FINISHPEG :STARTPEG :N-1)
END
```

The result returned by executing **SOLVE.HANOI "A "C "B 3** is the list of moves just described.

The **MK.MOVE** function is responsible for constructing each move. It has two inputs: the name of the start peg and the name of the destination peg. It returns a list containing a move.

```
TO MK.MOVE :STARTNAME :FINISHNAME
OP LIST LIST :STARTNAME :FINISHNAME
END
```

For example, **MK.MOVE "A "C** returns [[A C]].

To make things a little nicer, we could define a toplevel procedure that requires the user to supply only the value of :n.

```
TO HANOI :N
OP SOLVE.HANOI "A "C "B :N
END
```

The value returned by **HANOI** is a collection of moves that purportedly provide a solution to the Towers of Hanoi problem. If you are convinced, skip the rest of this section. If not, let's write a program that checks the solution. The program is named **CHECK.HANOI**. It takes a list of moves and a number indicating the number of disks for which the moves are a solution.

CHECK.HANOI begins by building three towers, stacking the appropriate number of disks on tower A. The towers are bound to the local variables :A, :B, and :C. It then calls a function named **CHECK.MOVES** to perform the moves. **CHECK.MOVES** performs all legal moves in the order they occur in the list. If it discovers an illegal move, it does not perform any more moves and returns a list of unperformed moves. **CHECK.HANOI** then displays all unperformed moves along with the current state of the towers.

```
TO CHECK.HANOI :MOVES :N
(LOCAL "A "B "C)
MAKE "A MK.TOWER :N
MAKE "B MK.TOWER 0
MAKE "C MK.TOWER 0
SHOW CHECK.MOVES :MOVES
SHOW :A
SHOW :B
SHOW :C
END
```

Each disk is represented by a positive integer. The numerical value represents the size of the disk. Higher values represent larger disks. MK.TOWER is a function that takes a nonnegative number n as its input value and returns a list of integers from 1 to n. The list of numbers represents a tower of n disks with the first element of the list denoting the top disk. Here's the definition of MK.TOWER:

```
TO MK.TOWER :N
IF :N = 0 [OP [ ]] [OP LPUT :N MK.TOWER :N-1]
END
```

CHECK.MOVES first determines if there are any more moves to perform. If not, it returns the empty list. If the list of moves is not empty, it examines the next one for legality. If the next move is illegal, the remaining moves are returned. If it is a legal move, the appropriate towers are modified and the remaining moves are checked.

```
TO CHECK.MOVES :MOVELIST
IF EMPTYP :MOVELIST [OP [ ]]
IF BAD.MOVE FIRST :MOVELIST [OP :MOVELIST]
PERFORM.MOVE FIRST :MOVELIST
OP CHECK.MOVES BF :MOVELIST
END
```

BAD.MOVE takes a move as its input and determines its legality by first determining if there are any disks on the originating tower. If there are none, the move is illegal. If the originating tower is not empty, it checks the destination tower. If there are no disks, the move is legal. If there are disks on both towers, it compares the size of the disk to be moved with the size of the disk on the top of the destination tower:

```
TO BAD.MOVE :MOVE
IF EMPTYTOWERP FIRST :MOVE [OP "TRUE]
IF EMPTYTOWERP LAST :MOVE [OP "FALSE]
  [OP GET.TOPDISK FIRST :MOVE >
      GET.TOPDISK LAST :MOVE]
END
```

GET.TOPDISK takes the name of a tower as its input. It gets the stack of disks (the list of numbers) associated with that tower and returns the top (first) one.

```
TO GET.TOPDISK :TOWERNAME
OP FIRST THING :TOWERNAME
END
```

PERFORM.MOVE saves the names of the towers involved in the move by binding them to the local variables **:FROMTOWER** and **:TOTOWER**. It then adds the disk to the destination tower and removes it from the originating tower. This order must be maintained. It we first modify the originating tower, we lose the disk, and we cannot add it to the destination tower.

```
TO PERFORM.MOVE :MOVE
LOCAL "FROMTOWER "TOTOWER
MAKE "FROMTOWER FIRST :MOVE
MAKE "TOTOWER LAST :MOVE
MAKE :TOTOWER FPUT GET.TOPDISK :FROMTOWER
                   THING :TOTOWER
MAKE :FROMTOWER BF THING :FROMTOWER
END
```

† **Exercise 7.8.1.** Modify the definition of **HANOI** so that it checks the solution as soon as the solution is generated.

Exercise 7.8.2. Define a function named **SOLVE.HANOIP** that takes a number n and a purported solution to the Towers of Hanoi problem as inputs. It returns **TRUE** if the solution is correct and **FALSE** otherwise.

7.9 Graphical Solution to the Towers of Hanoi Problem

The programs for solving the Towers of Hanoi problem given in the previous section seem to work, but they are not very satisfying because we cannot see the moves as they are happening. In this section we outline a graphical solution to the problem. That is, we define a program that depicts the actual solution graphically. We want to display the towers and disks and watch the disks moving from peg to peg until the solution is obtained. As a byproduct of this venture, we can easily define a program that allows the user to try solving the problem graphically.

Our first goal is to display the three towers in their initial configuration and define a move procedure that will allow the user to move the top disk of one tower to another tower. Each time the user enters a legal move, a disk will float from one tower to another.

MOVE takes two inputs: the names of the two towers involved in a single move. The disk on top of the tower named by the first input will be moved onto the tower named by the second input. For example, **MOVE** "A "C causes the disk on top of tower A to move to the top of tower C.

We treat the disks as graphical objects. Since the solution involves moving objects (disks) around the screen, we set up a world of disks that can be moved around by issuing appropriate requests.

Each disk has a location and a size. The location is a point at the center of the bottom of the disk. The size is represented by a number describing the radius of the disk. All disks have the same height, 5 units. We view the disks from the side, which means the disk is drawn as a rectangle. Thus, a disk with size 15 located at the point [100 0] is described in Figure 7.3.

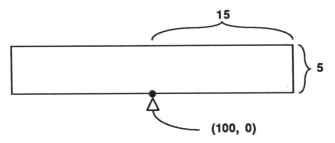

Figure 7.3 Picture of a Disk

Graphical Solution to the Towers of Hanoi Problem 251

A procedure named **MK.DISK** is used to create disks. **MK.DISK** has three inputs: a word used to name the disk, the location of the disk, and the size of the disk. It uses property lists to store the properties and is defined as follows.

Disk Constructor

MK.DISK takes a name, a location, and a size as inputs. It constructs a disk by adding location and size properties to its name. Since disks are always visible, it draws the disk on the screen.

```
TO MK.DISK :NAME :LOC :SIZE
PPROP :NAME "LOCATION :LOC
PPROP :NAME "SIZE :SIZE
DRAWDISK :LOC :SIZE
END

TO DRAWDISK :LOC :SIZE
DISKDRAWINIT :LOC
PENDOWN
DRAWREC :SIZE
END

TO DISKDRAWINIT :LOC
PU SETPOS :LOC SETHEADING 270
END

TO DRAWREC :SIZE
PENDOWN
FD :SIZE RT 90
FD 4 RT 90
FD 2*:SIZE RT 90
FD 4 RT 90 FD :SIZE
END
```

The next task is to define the functions that return the location and size of a particular disk as well as a procedure to move a disk to a new location. These are described and defined next.

Disk Functions

DISK.LOC *name*
DISK.LOC is a function that returns the location of the named disk.

```
TO DISK.LOC :NAME
OP GPROP :NAME "LOCATION
END
```

DISK.SIZE *name*
DISK.SIZE is a function that returns the size of the named disk.

```
TO DISK.SIZE :NAME
OP GPROP :NAME "SIZE
END
```

Disk-Move Procedure

MOVEDISK *name loc*
MOVEDISK is a procedure that moves the named disk to the indicated location.

It performs the move by

1. Erasing the current display of the disk.
2. Constructing the new representation of the disk.

Note: **MK.DISK** draws the picture of the disk as well as constructing the new representation.

```
TO MOVEDISK :NAME :NEWLOC
ERASEDISK DISK.LOC :NAME DISK.SIZE :NAME
MK.DISK :NAME :NEWLOC DISK.SIZE :NAME
END
```

```
TO ERASEDISK :LOC :SIZE
DISKDRAWINIT :LOC
PENERASE
DRAWREC :SIZE
END
```

We have now defined procedures that can move disks around, but since moves are specified in the form

"Move the top disk on tower A to tower C"

we need to determine what disk is on the top of tower A and what location denotes the top of tower C.

The solution to this problem is to view a tower as an object with two properties: a location and a stack of disks. Then we can extract the "top disk" from the appropriate tower's stack. We need a procedure to construct a tower and functions to select the values of its properties. The tower constructor and the selectors are defined as follows.

Tower Constructor

```
TO MK.TOWER :NAME :XLOC :N
PPROP :NAME "LOCATION :XLOC
PPROP :NAME "DISKS MK.DISK.NAMES :N
MK.DISK.STACK :N :XLOC -75
DRAWTOWER :NAME :XLOC
END

        TO MK.DISK.STACK :N :XLOC :YLOC
        IF :N < 1 [STOP]
          [ MK.DISK WORD "DISK :N
                    LIST :XLOC :YLOC
                    5*:N
            MK.DISK.STACK :N-1 :XLOC :YLOC+5]
        END
```

```
TO MK.DISK.NAMES :N
IF :N < 1 [OP [ ]]
   [OP FPUT WORD "DISK :N
              MK. DISK. NAMES :N - 1]
END
```

Tower Functions

TOWER.LOC *towername*
TOWER.LOC is a function that returns the location of the base of the indicated tower.

```
TO TOWER.LOC :NAME
OP GPROP :NAME "LOCATION
END
```

TOWER.DISKS *towername*
TOWER.DISKS is a function that returns the stack of disks on the indicated tower.

```
TO TOWER.DISKS :NAME
OP GPROP :NAME "DISKS
END
```

EMPTYTOWERP *towername*
EMPTYTOWERP is a function that returns **TRUE** if the indicated tower contains no disks and **FALSE** otherwise.

```
TO EMPTYTOWERP :NAME
OP EMPTYP TOWER.DISKS :NAME
END
```

MK.TOWER takes a name, the *x*-coordinate of its base and a number describing how many disks are to be placed on the tower. The *y*-

coordinate of the base of the tower is assumed to be -75 for each tower constructed. The location is saved on the property list of the tower. Then **MK.DISK.NAMES** is called to create and return a list of names of the disks to be associated with the tower. The first name in the list is the name of the top disk, the second name is the name of the second disk on the stack, and so on. The list is saved on the property list of the tower. **MK.DISK.STACK** creates the disks. Finally, the tower is drawn (see Exercise 7.9.1).

The n disks are constructed by **MK.DISK**. The smallest disk will have radius 5, the next smallest will have radius 10, and so on. The names of the disks constructed are (smallest to largest): **DISK1, DISK2,** ....

When we move a disk from the top of one tower to the top of another tower, we need to extract the name of the disk on top of the originating tower and determine the location of the top of the second tower. We define the needed operations next.

Additional Tower Functions

TOP.DISK *towername*
TOP.DISK is a function that returns the name of the disk on top of the indicated tower.

```
TO TOP.DISK :NAME
OP FIRST TOWER.DISKS :NAME
END
```

TOWER.TOP *towername*
TOWER.TOP is a function that returns the location of the top of the indicated tower. This is the location at which a disk can be added to the tower.

```
TO TOWER.TOP :NAME
OP LIST TOWER.LOC :NAME
        -75 + 5 * LENGTH TOWER.DISKS :NAME
END
```

TOP.DISK takes the name of a tower as input and returns the first element of the list representing the disk stack. **TOWER.TOP** determines the number of disks on its stack and multiplies the result by 5 (the height of a disk) to find the *y*-coordinate relative to the base of the tower. This number is added to the absolute coordinate of the base to get the absolute *y*-coordinate of the top of the tower. The *x*-coordinate is the tower location saved on the tower's property list and is obtained by the **TOWER.LOC** primitive.

We are now ready to define the procedure that actually performs the move. **MOVE** will take the names of two towers as inputs and move the top disk from the first input tower and place it on the top of the second tower. It is described next.

Moving Top Disk From One Tower To Another

MOVE.TOP.DISK *towername towername*
MOVE.TOP.DISK moves the topmost disk from the first tower onto the top of the second tower. If there is no disk on the originating tower, no action is performed.

```
TO MOVE.TOP.DISK :FROMTOWER :TOTOWER
LOCAL "NAME
IF EMPTYTOWERP :FROMTOWER [STOP]
MAKE "NAME TOP.DISK :FROMTOWER
MOVEDISK :NAME LIST TOWER.LOC :FROMTOWER 100
MOVEDISK :NAME LIST TOWER.LOC :TOTOWER 100
MOVEDISK :NAME TOWER.TOP :TOTOWER
POP.DISK :FROMTOWER
ADD.DISK :NAME :TOTOWER
END
```

MOVE gets the name of the top disk on the *from* tower and moves it to the *to* tower in three steps:

1. The disk is raised vertically to a position at the top of the screen directly over the *from* tower.

2. The disk is moved horizontally to a position directly over the *to* tower.

Graphical Solution to the Towers of Hanoi Problem

3. The disk is lowered vertically to the top position of the *to* tower.

After the disk has been moved, we must update the properties of the *from* tower and the *to* tower to reflect the new state. **POP.DISK** removes the top disk from the list of disks on the property list of a tower. **ADD.DISK** adds a new disk name onto the top of the disk stack on the property list of the *to* tower. Their definitions are

```
TO POP.DISK :TOWERNAME
   PPROP :TOWERNAME
         "DISKS
         BF TOWER.DISKS :TOWERNAME
END


TO ADD.DISK :DISKNAME :TOWERNAME
   PPROP :TOWERNAME
         "DISKS
         FPUT :DISKNAME TOWER.DISKS :TOWERNAME
END
```

† **Exercise 7.9.1.** Define **DRAWTOWER**. It should draw the tower along with its stack of disks on the display screen.

Exercise 7.9.2. Define a program named **MK.TOWERS.OF.HANOI**, which takes a positive integer n and creates the towers and disks and allows the user to manipulate the disks according to the rules of the game.

† **Exercise 7.9.3.** Define a program named **SOLVE.HANOI.GRAPHICALLY** that takes a positive integer n as its only input. It creates the towers and disks and solves the problem graphically by moving the pictures of the disks from tower to tower according to the rules of the game until the desired result is achieved.

Exercise 7.9.4. Modify **MOVE.TOP.DISK** so that nothing happens if the move results in a disk being placed on top of a smaller disk.

Exercise 7.9.5. Write a program that allows the user to try his or her own skill in solving the Towers of Hanoi problem. The three towers and a stack of disks are displayed on the screen. The user is then prompted for two tower names, indicating a move. The program shows the results of the move graphically if it is a legal move and complains if the move is an illegal one. The program also congratulates the user if he or she attains a solution.

Chapter 8
Building Your Own Computational Environment

8.1 Introduction

In this chapter we describe a framework that can be used to implement a variety of computational environments. It is illustrated by defining a rational number calculator in Logo.

The rational number calculator is a Logo program that prompts the user for an expression involving rational numbers, evaluates the expression, and displays the result. It works like this:

```
RATCALC> 1/2 + 1/4 * 1/3
7/12
RATCALC> 2/3 + 4/3 + 1
3
RATCALC>
```

When writing a program that defines a new computational environment, we need to insulate it from the behavior of the language we are using to define it. If something goes wrong during the use of the rational number calculator, we do not want the user to be staring at a Logo error message and the ? prompt. We would expect an error message and a new **RATCALC>** prompt. To build in the appropriate insulation, we first introduce a new type of control mechanism provided by Logo.

The control mechanisms that we have been using so far in writing Logo programs are

1. Sequencing (executing a sequence of statements).
2. Conditional branching (**IF**).
3. Procedure and function calling (includes recursive calling).
4. Returning from a procedure or function call (**OUTPUT** and **STOP**).

The new control mechanism that is introduced in this chapter is embodied by the Logo primitives named **CATCH** and **THROW**. It is an abnormal form of returning from a function or a procedure. We refer to this type of control as

    5. Escaping from a procedure or function call (**CATCH** and **THROW**).

It differs from normal returns (**OUTPUT** and **STOP**) in that control is not necessarily transferred back to the calling procedure. It is introduced in the context of solving a problem encountered in the **EXPLORE.GRAPH** program defined in Chapter 4.

8.2 The Graceful Return of EXPLORE.GRAPH

The **EXPLORE.GRAPH** program given next (and discussed in Chapter 4) has an annoying characteristic that we would like to change. Namely, the program doesn't terminate.

```
TO EXPLORE.GRAPH  :LOW   :HIGH   :INCREMENT
WINDOW
DRAW.GRAPH GEN.POINTS  :LOW  :HIGH  :INCREMENT
END

TO DRAW.GRAPH  :POINTS
CLEARSCREEN
CONNECT.POINTS  :POINTS
DRAW.GRAPH MODIFY.GRAPH  :POINTS  READCHAR
END

TO MODIFY.GRAPH  :POINTS :CHR
IF :CHR = "D [OP MOVEDOWN :POINTS  40]
IF :CHR = "U [OP MOVEUP :POINTS  40]
IF :CHR = "L [OP MOVELEFT :POINTS  40]
IF :CHR = "R [OP MOVERIGHT :POINTS· 40]
    [OP MODIFY.GRAPH  :POINTS  READCHAR]
END
```

The program modifies the points of the graph and redraws it according to what characters are entered by the user. The only way to terminate the program is by entering an escape character (CONTROL-G in Apple Logo) when we are through playing with it. This has the undesirable side effect of leaving the screen in an unclean state as well as preventing us from using the graphing program as a subprogram in a larger application.

In this section we illustrate how to termininate the exploration gracefully. The graceful termination is performed by redefining the program so that all procedures and functions return when the user indicates he or she is done exploring the graph.

We need to add a condition that allows the program to return to Logo's toplevel. Furthermore, the question of when the program should return should be in the user's hands. This implies that **MODIFY.GRAPH** (the part of the program that interacts with the user) should look for the proper signal.

We first decide that **EXPLORE.GRAPH** should be terminated if the user types a **Q**. Thus, we need to add a line to **MODIFY.GRAPH** that checks this event:

    IF :CHR = "Q [...]

But, what should **MODIFY.GRAPH** do when the **Q** key is pressed? The **STOP** operation is the wrong thing to put in the brackets. **MODIFY.GRAPH** was designed to return a value, namely, a list of points. Not returning anything will cause the calling program to issue an error message.

MODIFY.GRAPH needs to return a special list that **DRAW.GRAPH** will recognize as a signal to return. For example, we could return a list containing the word **DONE**. Then, **DRAW.GRAPH** could check for this occurrence. When it appears, **DRAW.GRAPH** will then clean up the screen and perform a normal return to **EXPLORE.GRAPH**. **EXPLORE.GRAPH** will then return normally to the toplevel or to the computation that called it. Using this scheme the definitions of **DRAW.GRAPH** and **MODIFY.GRAPH** become the graceful **EXPLORE.GRAPH** program:

```
TO  EXPLORE.GRAPH   :LOW  :HIGH  :INCREMENT
WINDOW
DRAW.GRAPH   GEN.POINTS :LOW  :HIGH   :INCREMENT
END

TO  DRAW.GRAPH    :POINTS
IF  :POINTS  =  [DONE]    [CLEANUP STOP]
CLEARSCREEN
CONNECT.POINTS   :POINTS
DRAW.GRAPH   MODIFY.GRAPH   :POINTS   READCHAR
END

TO  MODIFY.GRAPH    :POINTS  :CHR
IF  :CHR  =  "D [OP  MOVEDOWN :POINTS  40]
IF  :CHR  =  "U [OP  MOVEUP :POINTS  40]
IF  :CHR  =  "L [OP  MOVELEFT :POINTS  40]
IF  :CHR  =  "R [OP  MOVERIGHT :POINTS  40]
IF  :CHR  =  "Q [OP  [DONE] ]
    [OP  MODIFY.GRAPH    :POINTS   READCHAR]
END
```

The definition of **CLEANUP** is

```
TO CLEANUP
CLEARSCREEN
TEXTSCREEN
END
```

The Graceful Return of explore.graph 263

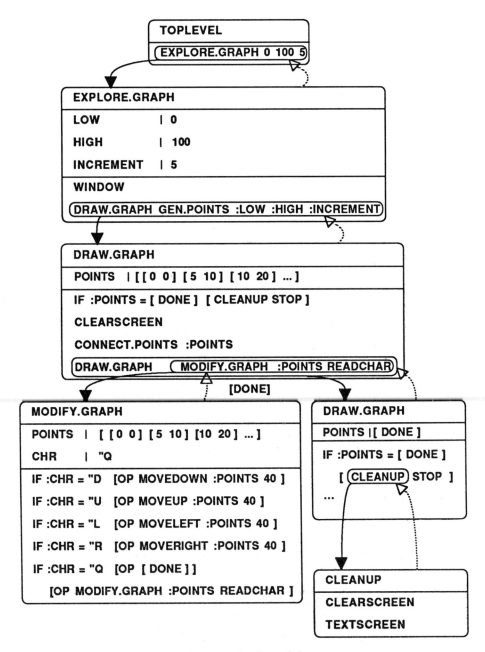

Figure 8.1 Executing the Graceful EXPLORE.GRAPH

Figure 8.1 describes the execution of **EXPLORE.GRAPH 0 100 5**. After obtaining the list of points from **GEN.POINTS**, **DRAW.GRAPH** is called. **DRAW.GRAPH** calls **CONNECT.POINTS** to draw the lines between the points and then calls **MODIFY.GRAPH**.

The level-three frame of **MODIFY.GRAPH** indicates that the user pressed the Q key. This causes it to return the list [DONE] to **DRAW.GRAPH** at level-two.

The level-three frame of **DRAW.GRAPH** is then created, which invokes **CLEANUP**. When **CLEANUP** returns, the level-three frame returns to the level-two frame of **DRAW.GRAPH**. The level-two frame has nothing more to do, so it returns to **EXPLORE.GRAPH** at level one. It, too, is finished, so it returns to the toplevel frame.

This new version of **DRAW.GRAPH** provides a clean way of terminating the exploration of the graph, but it too has its undesirable features. The solution required us to modify two procedures. Although this is not an inordinate amount of work, more complex programs may have a longer chain of returns to handle. The amount of work involved in getting a program to terminate gracefully depends on which procedure or function is responsible for determining that termination should occur and the intervening computations that are still active.

The second objection is that **MODIFY.GRAPH** was designed to return a list of points. Having it return the list [DONE] is not consistent with the design.

8.3 Going Directly to the Top

The previous section described how gracefully to exit the **EXPLORE.GRAPH** program. This section describes a less graceful but a direct and effective way of performing the exit task. It also eliminates the objections raised at the end of the previous section.

The idea is that when the user is ready to quit, all the program needs to do is to tidy up a bit and return directly to the toplevel. This action can be performed by using the original **DRAW.GRAPH** procedure and a single change in **MODIFY.GRAPH**. That is, the only modification that needs to be done is in the procedure that first recognizes it is time to terminate the program. The relevant definitions for the "escaping" **EXPLORE.GRAPH** programs and the program itself are as follows.

```
TO  EXPLORE.GRAPH   :LOW   :HIGH   :INCREMENT
WINDOW
DRAW.GRAPH    GEN.POINTS :LOW  :HIGH :INCREMENT
END

TO  DRAW.GRAPH      :POINTS
CLEARSCREEN
CONNECT.POINTS   :POINTS
DRAW.GRAPH    MODIFY.GRAPH    :POINTS    READCHAR
END

TO  MODIFY.GRAPH  :POINTS  :CHR
IF  :CHR  =  "D  [OP  MOVEDOWN  :POINTS  40]
IF  :CHR  =  "U  [OP  MOVEUP  :POINTS  40]
IF  :CHR  =  "L  [OP  MOVELEFT :POINTS  40]
IF  :CHR  =  "R  [OP  MOVERIGHT :POINTS  40]
IF  :CHR  =  "Q  [CLEANUP  THROW    "TOPLEVEL]
     [OP  MODIFY.GRAPH    :POINTS    READCHAR]
END
```

Now when the user types a **Q**, the **CLEANUP** procedure is executed, followed by the execution of **THROW "TOPLEVEL**. **THROW** is a Logo primitive procedure that transfers control to another frame in the sequence of currently active frames. In this setting, the effect is to transfer control directly to the toplevel—that is, to dispense with the current toplevel computation and begin a new one.

> **THROW** *word*
> **THROW** is a procedure that transfers control to the statement following an "active" **CATCH** *word* [. . .] statement. If no active **CATCH** statement of this form exists, then an error occurs. (See the description of **CATCH** in the next section.)*

Figure 8.2 describes the effect of executing the escaping version of **EXPLORE.GRAPH 0 100 5**. Everything proceeds as before until the level-three incantation of **MODIFY.GRAPH**. The user has pressed the **Q** key, so **CLEANUP** is executed. After returning from **CLEANUP**, the call to **THROW** is executed. **THROW** does not return anything; it simply junks the current computation and passes control somewhere else. In this case, control is passed to toplevel. Since there is nothing else to do at the toplevel, Logo waits for the user to enter a new statement.

---

*(1) **THROW** behaves very differently from a "go to" statement. In particular, it cannot be used to form a "looping" construct among a sequence of statements.

(2) There are two **CATCH**es embedded in Logo's toplevel that are always active. They will catch **THROW**s of the form **THROW "TOPLEVEL** and **THROW "ERROR**. They are examined in more detail in Section 8.5.

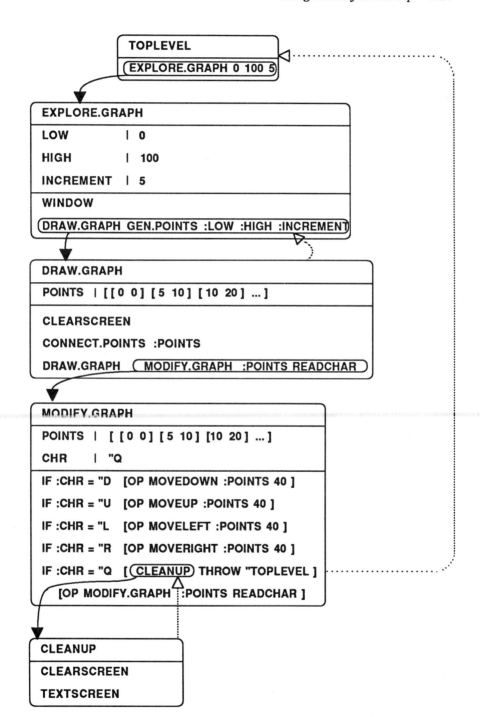

Figure 8.2 Executing the Escaping EXPLORE.GRAPH *Program*

The input to **THROW** may be any Logo word, but there are two special ones that are handled by Logo itself. One of these words is **TOPLEVEL** and the other is **ERROR**. We have described how **THROW "TOPLEVEL** is handled, and Section 8.5 describes how **THROW "ERROR** is dealt with. The next section illustrates how the programmer can modify the action of **THROW "TOPLEVEL** as well as the more general usage of **CATCH** and **THROW**.

Exercise 8.3.1. Can the definition of **DRAW.GRAPH** used in Section 8.2 be used with the definition of **MODIFY.GRAPH** given in this section? Explain.

8.4 Catching Control on the Way to the Top

The escaping version of the **EXPLORE.GRAPH** program avoided the objections raised about the graceful return version, but the escaping version has an even more serious objection. The escaping version cannot be used as a module in a larger program since the larger program, would also be dispensed with by the execution of **THROW "TOPLEVEL**. In the larger context we simply want **EXPLORE.GRAPH** to stop and return normally to the level that called it. That way the program that called it can proceed with another task. For example, consider the following program:

```
TO LARGE.PROGRAM
EXPLORE.GRAPH 0 100 5
DO.NEXT.TASK
END
```

The intent of **LARGE.PROGRAM** is first to explore the graph of some function. When the user decides to quit the exploration, the **DO.NEXT.TASK** procedure is executed. **EXPLORE.GRAPH** should return normally, so that the execution of **DO.NEXT.TASK** can be performed.

However, the escaping version of Section 8.3 causes control to be transferred to Logo's toplevel (above **LARGE.PROGRAM**), and **DO.NEXT.TASK** is never executed. **LARGE.PROGRAM** is dispensed with, along with **EXPLORE.GRAPH**.

Catching Control on the Way to the Top **269**

The Logo primitive **CATCH** provides us with the solution we need to modify the behavior of **LARGE.PROGRAM**. It can be used only inside a user-defined operation and is described next.

CATCH *catchword statementlist*
CATCH has two inputs: a word and a list of Logo statements. When executed, the **CATCH** is "activated" and the statements in *statementlist* are executed in the order given.

If during the execution of these statements, a **THROW** *throwword* is executed, then control is immediately transferred to the statement following the **CATCH** provided that

1. *catchword* = *throwword*, and

2. There is no other active **CATCH** *throwword* in the calling chain between the catch and the throw.

If all statements in *statementlist* are executed without executing such a **THROW**, then the statement immediately following the catch is executed. In either case the catch is deactivated prior to executing the statement following the catch statement.

If the statements in the statement list have been executed and no **THROW** is encountered, then the effect is the same as if the **CATCH** *catchword statementlist* were replaced by the sequence of statements in the list.

Logo's **Toplevel** program contains a **CATCH** "TOPLEVEL [...] (see Section 8.5), and that is why the **THROW** "TOPLEVEL works.

There may be more than one active **CATCH** with the same *catchword*. In that case, the **CATCH** "closest" to the corresponding **THROW** catches control. By closest, we mean the first active **CATCH** with the same throw word that we find by tracing the calling chain backward from the **THROW**. This fact allows us to redefine **LARGE.PROGRAM** to make everything work. If we defined **LARGE.PROGRAM1** as

```
TO LARGE.PROGRAM1
CATCH "TOPLEVEL [EXPLORE.GRAPH 0 100 5]
DO.NEXT.TASK
END
```

then the escaping **THROW "TOPLEVEL** that is executed inside **MODIFY.GRAPH** of Section 8.3 would be caught by **LARGE.PROGRAM1**, and **DO.NEXT.TASK** would then be executed. The level diagram in Figure 8.3 describes the behavior of the execution of **LARGE.PROGRAM1**.

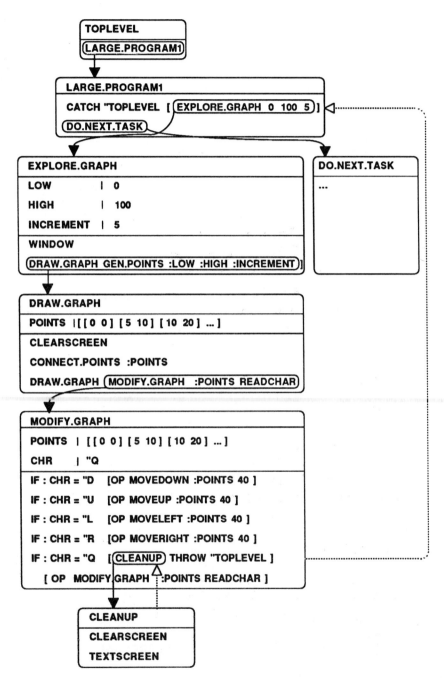

Figure 8.3 Catching the Escaping EXPLORE.GRAPH *Program*

If during the execution of **DO.NEXT.TASK**, another **THROW** "**TOPLEVEL** were executed, **LARGE.PROGRAM1** would not intercept the escaping **THROW**, since its **CATCH** is not "active" during the execution of **DO.NEXT.TASK**. Logo's **Toplevel** would catch such a **THROW** (provided that there is no other active **CATCH** between **LARGE.PROGRAM1** and Logo's **Toplevel**).

There is still an objection to the escaping **EXPLORE.GRAPH** program. When defining **LARGE.PROGRAM1**, the programmer must know that **EXPLORE.GRAPH** does a **THROW** "**TOPLEVEL** in order to catch it. This is unfortunate. **EXPLORE.GRAPH** should be a complete entity in itself. When we are defining **LARGE.PROGRAM1**, we should need to know only how to call **EXPLORE.GRAPH** and that it returns normally. If we expect to use **EXPLORE.GRAPH** as a module in a larger program, then it should always return in a normal way.

The appropriate solution is to embed the catch for the escaping **THROW** inside **EXPLORE.GRAPH**. That is, we should redefine **EXPLORE.GRAPH** (and friends) so that it is self-contained. This is the self-contained escaping **EXPLORE.GRAPH** program:

```
TO   EXPLORE.GRAPH   :LOW   :HIGH   :INCREMENT
WINDOW
CATCH "DONE [DRAW.GRAPH GEN.POINTS :LOW :HIGH :INCREMENT
END

TO   DRAW.GRAPH     :POINTS
CLEARSCREEN
CONNECT.POINTS    :POINTS
DRAW.GRAPH     MODIFY.GRAPH     :POINTS     READCHAR
END

TO   MODIFY.GRAPH   :POINTS  :CHR
IF   :CHR  =  "D  [OP   MOVEDOWN  :POINTS   40]
IF   :CHR  =  "U  [OP   MOVEUP  :POINTS   40]
IF   :CHR  =  "L  [OP   MOVELEFT  :POINTS   40]
IF   :CHR  =  "R  [OP   MOVERIGHT  :POINTS   40]
IF   :CHR  =  "Q  [CLEANUP   THROW    "DONE]
     [OP   MODIFY.GRAPH    :POINTS    READCHAR]
END
```

This version of **EXPLORE.GRAPH** is terminated by a **THROW**, which is caught at the top level of **EXPLORE.GRAPH** rather than at Logo's **Toplevel**. Now we can use the original definition of **LARGE.PROGRAM**, which did not require us to know anything at all about the workings of **EXPLORE.GRAPH**. We also use the word **DONE** as the throw word instead of **TOPLEVEL**.

Figure 8.4 describes the execution of **LARGE.PROGRAM**, which calls the self-contained version of **EXPLORE.GRAPH**. **EXPLORE.GRAPH** then calls **GENPOINTS**. When **GENPOINTS** returns, **DRAW.GRAPH** is called, which connects the points and calls **MODIFY.GRAPH**.

Finally, at level four, the user types a **Q**, and it is time to escape. The screen is cleaned up and **THROW "DONE** is executed. Looking back through the calling chain for an active **CATCH "DONE**, we find it in the level-two frame of **EXPLORE.GRAPH**. Control is immediately transferred to that frame, the catch is deactivated, and the line immediately following the **CATCH** is executed. There is nothing there, which means that **EXPLORE.GRAPH** does a normal return to the level-one frame. From there, **DO.NEXT.TASK** is called.

We have no objections to this version of **EXPLORE.GRAPH**.

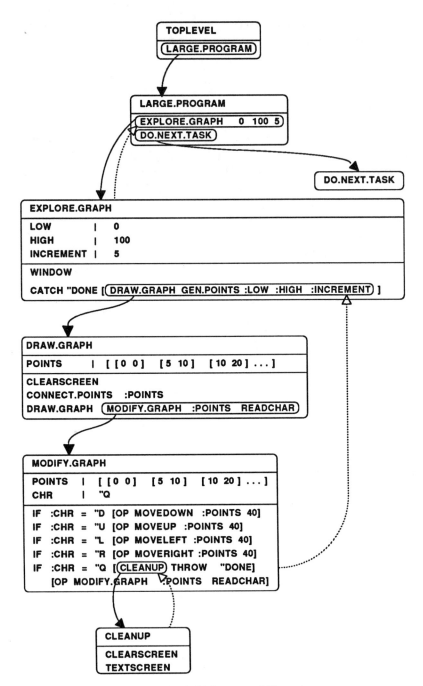

Figure 8.4 Executing the Self-Contained Escaping **EXPLORE.GRAPH**

8.5 Logo's Toplevel

In Section 2.2 the following program was presented as a description of how Logo works:

```
TO TOPLEVEL
TYPE "?
RUN READLIST
TOPLEVEL
END
```

We need to incorporate "escapes" to the toplevel into this description. That is, we need to add a **CATCH "TOPLEVEL [...]** to this program so that it will resume executing the basic loop described by **TOPLEVEL** when a **THROW "TOPLEVEL** is executed. Since we want to catch such throws while the user's input is executing, we wrap the catch around the **RUN READLIST** statement. The modified **TOPLEVEL** procedure looks like this:

```
TO TOPLEVEL
TYPE "?
CATCH "TOPLEVEL [RUN READLIST]
TOPLEVEL
END
```

Now, the call to **TOPLEVEL** in line three of the definition is executed if either of the following occurs:

1. A **THROW "TOPLEVEL** is executed during the execution of **RUN READLIST**.

2. The execution of **RUN READLIST** terminates normally.

There is one more part of Logo's **Toplevel** Module that is not currently described by our **TOPLEVEL** program, namely, error handling.

The **CATCH-THROW** mechanism is ideally suited to handle error situations. There are two ways that this mechanism could be used:

1. When an error is encountered, call a procedure that displays the appropriate error messages and then executes a **THROW "TOPLEVEL** statement.

276 *Building Your Own Computational Environment*

2. When an error is encountered, immediately transfer control to the **TOPLEVEL** program and have it handle the display of error messages.

Logo opts for the second solution to the error-handling problem. One good reason for doing this is that it allows the user to write programs that intercept Logo-generated errors. This feature is needed to insulate our programs from Logo error messages and is discussed in the next section.

When an error is detected Logo transfers control to the **TOPLEVEL** program by executing a **THROW "ERROR** statement. This means that somewhere in our **TOPLEVEL** program we need to catch such throws. After catching the throw to **ERROR**, the appropriate error message is displayed and the **TOPLEVEL** loop is reentered. We can do this by leaving the **TOPLEVEL** program alone and writing a higher-level procedure named **LOGO**:

```
TO LOGO
CATCH "ERROR [TOPLEVEL]
HANDLE. ERROR  ERROR
LOGO
END
```

The **LOGO** program simply calls the **TOPLEVEL** program. If a **THROW "ERROR** is executed anytime during the execution of **TOPLEVEL**, **HANDLE.ERROR** is called to display the appropriate error message. The **LOGO** program then invokes itself in order to reenter the **TOPLEVEL** program. Note that **TOPLEVEL** never returns. Thus, the only way **HANDLE.ERROR** is called is if a **THROW "ERROR** is executed during the execution of **TOPLEVEL**.

HANDLE.ERROR (not a Logo primitive) must obtain information about when and where the error occurred as well as why it occurred. Since **THROW** is a procedure and therefore does not return anything, the Logo primitive function named **ERROR** provides the necessary information. The assumption here is that the information is tucked away in a special place before the **THROW "ERROR** is executed. This special place is known by **ERROR**, which simply returns it.

> **ERROR**
> **ERROR** returns a list of six elements providing the information about the error. The elements are

> A number identifying the type of error.
> A message explaining the error.
> The name of the function or procedure in which the error occurred.
> The line where the error occurred.
> The name of the Logo primitive that generated the error.
> The object (if any) that caused the error.
>
> If no error has occurred since the last time **ERROR** was called, the empty list is returned.

ERROR also has the side effect of destroying the saved information. That is, once the error has been read by **ERROR**, the information about the error is thrown away. This is done because the user can execute a **THROW** "ERROR anytime he or she wishes and it would be misleading to report an error that has no bearing on the current situation.

As an example of what happens when an error occurs, suppose that the following statement is entered:

SHOW "MUMBLE + 1

The error condition is detected when the addition operation is called. That is, the definition of **+** checks to make sure its inputs are numbers. When it discovers that its first input is not a number, the type of error and the other information about why and where it occurred is recorded in the special area previously mentioned. Control is then transferred back to the **LOGO** program by executing **THROW** "ERROR. The error handler, **HANDLE.ERROR**, displays the appropriate message and the **Toplevel** program is then resumed.

8.6 The Logo Calculator

Since the information about an error is saved away and not displayed until control has passed to the very top, the user has a chance of intercepting the Logo error before anything gets displayed on the screen. To do this, the user simply includes a **CATCH "ERROR [. . .]** surrounding

Since the user's program is always at a lower level than the **CATCH "ERROR** embedded in Logo's **toplevel**, the user will intercept it.

An example of a program that we want to insulate from Logo-generated errors is one that defines a Logo calculator, a version of Logo that reads in Logo expressions and displays their values without having to enter a **SHOW** or **PRINT** statement. Such a program is defined as follows:

```
TO LOGOCALC
TYPE ">
CATCH "ERROR [SHOW RUN READLIST]
HANDLE.LOGO.ERROR ERROR
LOGOCALC
END
```

This program allows normal use of Logo with two slight differences. First, the user is prompted by the **>** character instead of the **?** character. Secondly, an error is generated whenever we enter a statement rather than an expression, because **SHOW** is not provided with an input. Such errors are caught and handled according to the definition of **HANDLE.LOGO.ERROR**. If the program did not catch Logo-generated errors, the program would be terminated as soon as a statement or an illegal expression was entered.

HANDLE.LOGO.ERROR ignores an error that results from trying to **SHOW** something that is not returned. This error is a type 10 error. All other errors are displayed. After handling the error, **LOGOCALC** is reentered. Note that **HANDLE.ERROR** is called whether or not an error occurs. **HANDLE.ERROR** can recognize this situation by looking at its input. If no error has occurred, then **ERROR** returns an empty list.

When writing applications programs, the program itself should be in control of errors. When a program detects an error condition, such as the wrong type of input to a procedure, an error message should be displayed and control passed to the appropriate place. Since the same type of error situation can arise in different places of the program, a centralized error handler is desired. That is, there should be a single procedure to insure a uniform handling of errors.

The error-handling mechanism can be an elaborate one similar to that of Logo, but for most applications a simpler one suffices. When an error condition is discovered by some part of a program, the general error-handler should be called with an error message and the object (if there is handler should be called with an error message and the object (if there is one) that caused the problem as inputs. It simply displays the error message and the object causing the problem and then passes control to wherever program execution should continue. Such an error handler could be defined as follows:

```
TO GENERROR :MSG :OBJ
PRINT :MSG
PRINT :OBJ
THROW "CONTINUE
END
```

After displaying the information on the screen, **GENERROR** passes control by escaping to a continuation point that picks up the normal execution of the program.

Exercise 8.6.1. Define **HANDLE.LOGO.ERROR** as just described.

Exercise 8.6.2. Define a program named **DEBUG**, which has no inputs and does the following:

a. Reads in a statement from the user and executes it.
b. If a Logo error occurs the following happens:
  (i) The error information is displayed.
  (ii) The user is asked if he or she wants to edit a function or procedure.
  (iii) If the user responds by entering the name of a procedure, the editor is entered, displaying the definition of that procedure. When the editor is exited, the statement entered by the user in (a) is executed again and the steps in (b) are performed again.
  (iv) If the user responds by pressing the RETURN key, Steps (a), (b) and (c) are performed again.
c. If no error occurs, Steps (a), (b) and (c) are performed again.

8.7 Rational-Number Calculator

A rational-number calculator should read and evaluate expressions that include integers and fractions written as the quotient of two integers. The operations performed are addition, subtraction, multiplication, and division of fractions. The result should be displayed as an integer or as a quotient of two integers. For example, a typical interaction with the rational-number calculator is as follows:

```
RATCALC> 1/2 + 1/4
3/4
RATCALC> 1 + 1/3 * 1/2
7/6
RATCALC> 1 + 1/3 + 2/3
2
```

We begin by defining a top-level procedure name, **RATCALC**, which displays the **RATCALC>** prompt, reads in an expression, evaluates (runs) it, displays the results, and then calls itself to perform these actions again. We enclose it in a higher-level procedure name **RAT** to catch any Logo-generated errors.

```
TO RAT
CATCH "ERROR [RATCALC]
HANDLE.ERRORS  ERROR
RAT
END

TO RATCALC
TYPE "RATCALC>
RATPRINT RATRUN RATREAD
RATCALC
END

TO HANDLE.ERRORS :ERRLIST
IF 28 = ITEM 1 :ERRLIST [THROW "TOPLEVEL]
IF NOT EMPTYP :ERRLIST [PRINT FIRST BF :ERRLIST]
END
```

The first line of **HANDLE.ERRORS** enables the program to terminate when a CNTRL-G is entered. CNTRL-G causes a type 28 error. If this type of error occurs, we want to return control immediately to Toplevel. Without this line, the program cannot be terminated except by turning off the computer.

Logo-generated errors are handled by printing the error message. The other information contained in **:ERRLIST** is suppressed.

The program detects other kinds of errors that are specific to the rational-number calculator. When that happens, a procedure named **GEN.RAT.ERROR** is called. Its definition is given by

```
TO GEN.RAT.ERROR :MSG :OBJ
PRINT :MSG
PRINT :OBJ
THROW "ERROR
END
```

The **THROW "ERROR** statement causes control to be passed to **RAT** where **HANDLE.ERRORS** is executed. Thus, Logo-generated errors as well as program-generated errors go to the same place. **HANDLE.ERRORS** is called in both instances, but it does not do anything if a program-generated error has occurred (**:ERRLIST** is empty).

Reading and evaluating a rational-number expression are handled by the Logo programs **RATREAD** and **RATRUN**, respectively. Printing the result (a rational number) is performed by the Logo program named **RATPRINT**.

RATREAD reads the string of characters entered by the user and returns the Logo representation of the expression. **RATRUN** takes the representation returned by **RATREAD** and simplifies it to the representation of a rational number and returns it. We begin by describing what we mean by a rational-number expression and define the simplifier.

8.8 Ratrun: The Rational-Number Expression Simplifier

Every rational-number expression can be viewed as either a rational number constant, a sum of two rational expressions, a difference of two rational expressions, a product of two rational expressions, or a quotient of two rational expressions. This description can be compactly summarized using BNF (Backus Naur form) notation. BNF is used to define the structure of an object and was invented in 1960 to define the structure of the ALGOL programming language.

A definition in BNF consists of a single object to the left of the ::= symbol and a description of the object to the right of the ::= symbol. Angle brackets are used to denote an object.

A rational-number expression is defined next. The form <rat exp> denotes a rational-number expression.

Definition of a Rational Expression

<rat exp> ::= <constant> | <rat exp> <oper> <rat exp>
<oper> ::= + | - | * | /
<constant> ::= <integer> | <integer> <slash> <integer>
<slash> ::= /

We read this notation by substituting the following English words for the occurrence of the symbols, ::=, |, and > <.

The symbol ::= is read "is defined to be."
The symbol | is read "or a."
The combination > < is read "followed by."

Using this translation scheme, the first line in the definition says,

A rational expression is defined to be a constant or a rational-number expression followed by an operation followed by another rational-number expression.

Note that <oper> denotes one of: +, -, *, /.

Assuming that we have functions **RAT.CONSTANTP, RAT.SUMP, RAT.DIFFP, RAT.TIMESP,** and **RAT.DIVIDEP** that will recognize the kind of rational-number expression that was entered, we can begin to define **RATRUN.**

RATRUN takes a single input, namely, the representation of a rational-number expression. It looks at it and and decides what to do in each case. The skeleton of **RATRUN** is:

```
TO RATRUN :RATEXP
IF RAT.CONSTANTP :RATEXP [ ... ]
IF RAT.SUMP :RATEXP [ ... ]
IF RAT.DIFFP :RATEXP [ ... ]
IF RAT.PRODUCTP :RATEXP [ ... ]
IF RAT.QUOTIENTP :RATEXP [ ... ]
GEN.RAT.ERROR [UNRECOGNIZABLE EXPRESSION:] :RATEXP
END
```

If **:RATEXP** is a constant, **RATRUN** simply returns it. In the other cases, **RATRUN** must simplify the input expressions to the operation and then call on the appropriate function to return a representation of the result.

The definition of **RATRUN** can now be written as

```
TO RATRUN :RATEXP
IF RAT.CONSTANTP :RATEXP [OP :RATEXP]
IF RAT.SUMP :RATEXP
   [OP RATPLUS   RATRUN ARG1 :RATEXP
                 RATRUN ARG2 :RATEXP]
IF RAT.DIFFP :RATEXP
   [OP RATMINUS RATRUN ARG1 :RATEXP
                RATRUN ARG2 :RATEXP]
IF RAT.PRODUCTP :RATEXP
   [OP RATTIMES RATRUN ARG1 :RATEXP
                RATRUN ARG2 :RATEXP]
IF RAT.QUOTIENTP :RATEXP
   [OP RATDIVIDE RATRUN ARG1 :RATEXP
                 RATRUN ARG2 :RATEXP]
GEN.RAT.ERROR [UNRECOGNIZABLE EXPRESSION:] :RATEXP
END
```

The value returned by **RATRUN ARG1 :RATEXP** and **RATRUN ARG2 :RATEXP** is the representation of a rational number. Likewise, the results returned by **RATPLUS, RATMINUS, RATTIMES,** and **RATDIVIDE** are representations of rational numbers.

As yet we have said nothing about the actual representation of a rational-number expression. The definitions of the recognizers

284 *Building Your Own Computational Environment*

(**RAT.CONSTANTP RAT.SUMP, RAT.DIFFP, RAT.PRODUCTP**, and **RAT.QUOTIENTP**) and selectors (**ARG1**, and **ARG2**) depend on the representation, but the representation ultimately chosen will have no effect on the definition of **RATRUN** itself. At this time we just assume the existence of these selectors and recognizers for rational-number expressions.

Since the sum of two rational numbers a/b and c/d can be expressed by

$$a/b + c/d = (ad + bc)/bd \qquad (1)$$

we can define the function **RATPLUS**, again without choosing actual representations for the rational numbers. We need only assume the existence of selectors named **NUMERATOR** and **DENOMINATOR** and a constructor named **MK.RAT**.

Using (1), and these functions,

Constructor and Selectors for Rational Numbers

Constructor

MK.RAT *integer integer*
MK.RAT takes two integers as inputs and returns the representation of a rational number whose numerator is the first input and whose denominator is the second.

Selectors

NUMERATOR *rational number*
NUMERATOR returns an integer representing the numerator of the rational number.

DENOMINATOR *rational number*
DENOMINATOR returns an integer representing the denominator of the rational number.

---

we can define **RATPLUS** as follows:

```
TO RATPLUS :RAT1 :RAT2
  OP MK.RAT (NUMERATOR :RAT1) * (DENOMINATOR :RAT2)
            + (NUMERATOR :RAT2) * DENOMINATOR :RAT1
            (DENOMINATOR :RAT1) * DENOMINATOR :RAT2
END
```

RATRUN and **RATPLUS** are called representation-independent functions. Their definitions do not know and do not care how rational-number expressions and rational numbers are represented. We are free to select any representation we want.

Exercise 8.8.1. Define representation-independent functions named **RATMINUS**, **RATTIMES**, and **RATDIVIDE** that perform the subtraction, multiplication, and division of two rational numbers.

† **Exercise 8.8.2.** Define a representation-independent procedure named **RATPRINT**. It takes the representation of a rational number and displays it on the screen. Rational numbers that are integers should be displayed as integers.

8.9 The Representation of Rational-Number Expressions

None of the representation-independent functions defined in the previous section are useful until we define all the selectors, constructors, and recognizers whose existence was assumed. We can not define any of these until we choose a representation for rational-number expressions.

The operations we need to define are given next.

Recognizers and Selectors for Rational-Number Expressions

Recognizers

RAT.CONSTANTP *rational-numberexpression*
RAT.CONSTANTP returns **TRUE** if the expression is the representation of a rational number and returns **FALSE** otherwise.

> **RAT.SUMP** *rational-numberexpression*
> **RAT.SUMP** returns **TRUE** if the expression is the representation of the sum of two rational-number expressions, and returns **FALSE** otherwise.
>
> **RAT.DIFFP** *rational-numberexpression*
> **RAT.DIFFP** returns **TRUE** if the expression is the representation of the difference of two rational-number expressions, and returns **FALSE** otherwise.
>
> **RAT.PRODUCTP** *rational-numberexpression*
> **RAT.PRODUCTP** returns **TRUE** if the expression is the representation of the product of two rational-number expressions, and returns **FALSE** otherwise.
>
> **RAT.QUOTIENTP** *rational-numberexpression*
> **RAT.QUOTIENTP** returns **TRUE** if the expression is the representation of the quotient of two rational-number expressions, and returns **FALSE** otherwise.
>
> *Selectors*
>
> **ARG1** *combination*
> **ARG1** returns the first argument to the operation in the specified combination.
>
> **ARG2** *combination*
> **ARG2** returns the second argument to the operation in the specified combination.

Recalling the definition of a rational-number expression given in Section 8.8,

<rat exp> ::= <constant> | <rat exp> <oper> <rat exp>
<oper> ::= + | - | * | /
<constant> ::= <integer> | <integer> <slash> <integer>
<slash> ::= /

we see that there are two types: constants and combinations. Constants have two components, an integer representing the numerator and an integer representing the denominator. Thus, we choose a list of two integers to represent each rational number. The first element represents the numerator and the second represents the denominator. We use the following notation to indicate this choice:

<constant> --> [<numerator> <denominator>]

That is,

The list **[2 3]** represents the rational number 2/3.

The list **[5 1]** represents the rational number 5.

Using this representation, we can define **MK.RAT, NUMERATOR, DENOMINATOR,** and **RAT.CONSTANTP** as follows:

```
TO MK.RAT :N :D
OP LIST :N :D
END

TO NUMERATOR :RAT
OP FIRST :RAT
END

TO DENOMINATOR :RAT
OP LAST :RAT
END

TO RAT.CONSTANTP :OBJ
IF NOT LISTP :OBJ [OP "FALSE]
IF (LENGTH :OBJ) = 2
   [OP AND NUMBERP FIRST :OBJ NUMBERP LAST :OBJ]
   [OP "FALSE]
END
```

Combinations have three components: an operation and two rational-number expressions. We can represent combinations as a three-element list:

<rat exp> <oper> <rat exp> --> [<oper> <rat exp> <rat exp>]

Each operation is represented by itself:

```
+ ---> +
- ---> -
* ---> *
/ ---> /
```

The following are some examples of the representations of some rational-number combinations:

```
1/2 + 3/4       --->   [+ [1 2] [3 4]]
1/2 + 2/3 * 5   --->   [+ [1 2] [* [2 3] [5 1]]]
2 * 3/5 + 1/3 * 2/7 ---> [+ [* [2 1] [3 5]] [* [1 3] [2 7]]]
```

Exercise 8.9.1. Assuming the given representations, define each of the following operations:

† a. **RAT.SUMP**
  b. **RAT.DIFFP**
  c. **RAT.PRODUCTP**
  d. **RAT.QUOTIENTP**
† e. **ARG1**
  f. **ARG2**

† **Exercise 8.9.2.** The definition of **MK.RAT** returns an unsimplified rational number. Redefine **MK.RAT** so that it always returns a simplified rational number. That is, **MK.RAT 3 6** should return the list **[1 2]** rather than the list **[3 6]**).

To simplify a rational number we divide the numerator and denominator by the **greatest common divisor** (GCD). Thus, the first step should be to define a function that computes the GCD of two integers.

The following relationship, discovered by Euclid, provides an efficient way to compute the GCD:

$$GCD(a, b) = \begin{cases} a & \text{if } b = 0 \\ GCD(b, r) & \text{if } b \neq 0 \end{cases}$$

where r is the remainder obtained by dividing a by b.

8.10 Ratread: The Rational-Number Calculator Reader

The purpose of a reader is to read in a string of characters entered by the user and return a representation of the input that is appropriate for further processing. In the case of the rational-number calculator, the reader is named **RATREAD** and it returns a list representing a rational-number expression.

The reader consists of three interrelated tasks and three global variables. The variables are **:INPUTSTREAM**, **:TOKEN**, and **:OPTABLE**. The three main tasks are described by procedures named **READLINE**, **RATSCAN** and **RATPARSE**.

The **READLINE** procedure interacts with the user and sets the global variable **:INPUTSTREAM** to a word containing the characters entered by the user. For example, if the user enters 1/2 + 1/3 * 1/4, **READLINE** sets **:INPUTSTREAM** to the word 1/2 + 1/3 * 1/4. Note that spaces are included as characters in the word.

The **RATSCAN** procedure removes just enough characters from the value of **:INPUTSTREAM** to obtain one of the basic objects (called a token) in a rational expression and assign the token to **:TOKEN**. The tokens are the representations for rational numbers (a two element list), the operations, +, -, *, /, and an end-of-line token. The end-of-line token is represented by the word **EOL**. It is obtained if there are no more characters in **:INPUTSTREAM**. The successive tokens in the expression 1/2 + 1/3 * 1/4 are [1 2], +, [1 3], *, [1 4], and **EOL**. A procedure that obtains a single token from the string of characters describing the text entered by the user is called a **scanner**.

RATPARSE uses the tokens retrieved by **RATSCAN** to build the list representation of the rational-number expression. The list constructed depends on the binding powers of the arithmetic operations used in the expressions. In the above example **RATPARSE** returns the list

    [+ [1 2] [* [1 3] [1 4]]]

We use the global variable **:OPTABLE** to store a table of binding powers for each operation. The operation table is defined by executing

    MAKE "OPTABLE [[+ 5 6] [- 5 6] [* 7 8] [/ 7 8] [EOL 0 0]]

Note that the **EOL** token is regarded as an operation. This is done because a rational expression ends with a rational number and the token following a rational number is expected to be an operation. The binding powers of **EOL** are the lowest of all operations to ensure that we stop reading the expression when we encounter it.

Here is the definition of **RATREAD**:

```
TO RATREAD
MAKE "INPUTSTREAM READLINE
RATSCAN
OP RATPARSE :TOKEN 1
END
```

RATREAD initializes the global variable **:INPUTSTREAM**. The call to **RATSCAN** initializes **:TOKEN**. Note that the initialization of **:OPTABLE** is performed only once because it will never change, while **:INPUTSTREAM** is set every time that **RATREAD** is called.

READLINE returns the string of characters entered by the user. That is, the value of **:INPUTSTREAM** is a word. Spaces and other special characters (for example, **+**, *****, **-**, and **/**) are considered individual characters in the word making up the value of **:INPUTSTREAM**. The input line is terminated when the RETURN key is entered. Here is the definition of **READLINE**:

```
TO READLINE
OP READLINE.AUX READCHAR "
END
```

```
TO READLINE.AUX :CHR :STRING
TYPE :CHR
IF RETURNP :CHR
   [OP :STRING]
   [OP READLINE.AUX READCHAR WORD :STRING :CHR]
END
```

READLINE begins by calling an auxillary function named **READLINE.AUX** with the first character entered by the user as the first input and the empty word as the second input. The input variable named

:CHR is the most recent character read while the input variable named :STRING is used to accumulate the characters entered by the user.

READLINE.AUX displays the character read (otherwise, nothing will be displayed on the screen). It then checks to see if the user entered the RETURN character. If the RETURN character was pressed, the value of :STRING (the text entered by the user) is returned. Otherwise, **READLINE.AUX** is called again to read the remaining characters after appending the current character onto the end of :STRING.

The definition of **RETURNP** is given by:

```
TO RETURNP :CHR
OP (ASCII :CHR) = 13
END
```

It uses the Logo primitive function named **ASCII**. A character entered from the keyboard (a letter, digit, special characters, and control characters) is represented by an integer from 0 to 255. The first 128 code numbers refer to a standard code used by almost all computer manufacturers and include upper and lower case letters, digits, other visible characters such as $, +, and (, as well as control characters and certain invisible characters such as space and RETURN. This code for characters is called the ASCII code. The Logo primitive function, **ASCII**, returns the integer code for its character input. For example,

```
?SHOW ASCII "A
65
?SHOW ASCII "B
66
?SHOW ASCII "$
36
```

The ASCII code for the RETURN character is 13. The ASCII code for any character can be obtained by entering

```
?SHOW ASCII READCHAR
```

and then pressing the desired key. For example, if we press the space bar we get the response telling us that the ascii code for the space character is 32.

Exercise 8.10.1. Define an operation named **RAT.OPERP** that takes a word as its input and returns **TRUE** if its input denotes an operation that is in the operation table.

† **Exercise 8.10.2.** Define functions named **GET.LBP** and **GET.RBP**. Each has the symbol representing an operation as its input. **GET.LBP** returns the left binding power of its input, and **GET.RBP** returns the right binding power of its input. They should generate an error by calling **GEN.RAT.ERROR** if the operation is not in the table.

Exercise 8.10.3. Test the above definition of **READLINE** by entering

    ?SHOW READLINE

and then entering a sequence of characters from the keyboard followed by RETURN.

Exercise 8.10.4. The definition of **READLINE** is still not ideal because if a mistake is made, it does not allow the user to correct the error. To correct this problem, we can redefine **READLINE.AUX** as follows:

```
TO READLINE.AUX :CHR :STRING
IF DELP :CHR
  [ BACKSPACE
    OP READLINE.AUX READCHAR WORD :STRING :CHR]
TYPE :CHR
IF RETURNP :CHR
  [OP :STRING]
  [OP READLINE.AUX READCHAR WORD :STRING :CHR]
END
```

Define **DELP** and **BACKSPACE**. **DELP** recognizes the character to be used as the "delete previous character" key. **BACKSPACE** must erase the previous character from the screen and position the cursor so that the next character entered will appear at the proper location. Why is the **IF** statement placed before **TYPE :CHR**?

8.11 The Scanner

The job of the scanner, **RATSCAN**, is to set the value of the global variable **:TOKEN** to the next token in the input stream and to remove the characters describing the token from the front of the word stored in the global variable **:INPUTSTREAM**.

The tokens that will be returned are the following: the representation of a rational number, the word **EOL** (if the input stream is empty), or one of the symbols, **+**, **-**, *****, **/**. Each successive call puts a new value into **:TOKEN**. For example, if the user typed in

    1 + 1/3 * 1/2

the values of **:INPUTSTREAM** and **:TOKEN** would change as follows by successive calls to **RATSCAN**:

:INPUTSTREAM	:TOKEN
1 + 1/3 * 1/2	
+ 1/3 * 1/2	[1 1]
1/3 * 1/2	+
* 1/2	[1 3]
1/2	*
	[1 2]
	EOL

Further calls to **RATSCAN** would just keep setting **:TOKEN** to **EOL**. Spaces will be used as delimiters only. That is, they will signal the end of a token. After a space has served its delimiting purpose, it will be thrown away.

The definition of **RATSCAN** is as follows:

```
TO RATSCAN
MAKE "TOKEN SCAN PEEK.CHAR
END
```

That is, it sets the value of **:TOKEN** to the value returned by a function named **SCAN**.

PEEK.CHAR is a function that returns a copy of the next character in **:INPUTSTREAM**. If there are no characters in the inputstream, **PEEK.CHAR** will return the empty word:

```
TO PEEK.CHAR
IF EMPTYP :INPUTSTREAM
   [OP " ]
   [OP FIRST :INPUTSTREAM]
END
```

Note that **PEEK.CHAR** does not modify the value of **:INPUTSTREAM**.

SCAN begins by checking to see if there are any characters in **:INPUTSTREAM**. If not, the word **EOL** is returned as the next token. If there is at least one character, it checks to see if the character is an operation. If so, then the character is returned as the token after deleting the first character of **:INPUTSTREAM**. If the first character is a digit, then it is the beginning of a rational number and we call a function named **GET.RAT** to read it and return the representation for the number. If the first character is a space, it is removed (ignored) and we scan the character following the space.

Here is the definition of **SCAN**:

```
TO SCAN :NEXTCHR
IF EMPTYP :NEXTCHAR [OP "EOL ]
IF RAT.OPERP :NEXTCHAR [OP GET.CHAR]
IF DIGITP :NEXTCHAR [OP GET.RAT]
IF SPACEP :NEXTCHAR [REMOVE.CHAR OP SCAN PEEK.CHAR]
   [GEN.RAT.ERROR [ ILLEGAL CHARACTER IN INPUT:]
                  :NEXTCHAR]
END
```

GET.CHAR removes the first character from **:INPUTSTREAM** and returns a copy of that character, while **REMOVE.CHAR** simply removes the first character from **:INPUTSTREAM**. The definitions of **GET.CHAR** and **REMOVE.CHAR** are:

```
TO GET.CHAR
LOCAL "CHR
IF EMPTYP :INPUTSTREAM [OP " ]
   [ MAKE "CHR FIRST :INPUTSTREAM
     MAKE "INPUTSTREAM BF :INPUTSTREAM
```

```
    OP :CHR]
END

TO REMOVE.CHAR
IF NOT EMPTYP :INPUTSTREAM
    [MAKE "INPUTSTREAM BF :INPUTSTREAM]
END
```

The modification of :INPUTSTREAM is performed by GET.CHAR and REMOVE.CHAR only. No other procedures modify it except of course when it is initialized by RATREADLINE.

GET.RAT reads and returns the representation of a rational number. It gets the numerator and the denominator from the input stream and constructs the representation for a rational number:

```
TO GET.RAT
OP MK.RAT GET.NUMERATOR GET.DENOMINATOR
END
```

MK.RAT is given by

```
TO MK.RAT :NUM :DEN
OP LIST :NUM :DEN
END
```

GET.NUMERATOR reads the first digit of the numerator from the input stream and calls GETNUMBER to read the remaining digits:

```
TO GET.NUMERATOR
OP GETNUMBER GET.CHAR
END

TO GET.NUMBER :N
IF NOT NUMBERP :PEEKCHAR [OP :N]
    [OP GET.NUMBER WORD :N GET.CHAR]
END
```

GET.DENOMINATOR is a little more complex. It must determine if the input is a fraction or an integer. If it is a fraction, GET.DENOMINATOR returns the number representing the denominator. If the input is not a fraction, GET.DENOMINATOR will return 1.

It is clear what to do when scanning the expression **2 + 3/4**. The denominator associated with **2** is **1**, but the expression **2/3/4** is ambiguous. It can be viewed as **2** divided by **3/4** (the result is **8/3**) or as **2/3** divided by **4** (the result is **2/12**).

To resolve the ambiguity, we will assume that if the two characters immediately following the numerator are a slash and a digit, then the expression is a fraction. If a space or any other character is seen in the next two characters, then the denominator is assumed to be 1. Thus, each of the following expressions has the stated interpretation.

Expression	Interpretation
2+3/4	2 plus 3/4
2/3/4	2/3 divided by 4
2/ 3/4	2 divided by 3/4
2 /3/4	2 divided by 3/4
2 / 3/4	2 divided by 3/4

GET.DENOMINATOR is defined by

```
TO GET.DENOMINATOR
IF FRACTIONP
   [REMOVE.CHAR   OP GETNUMBER GET.CHAR]
   [OP 1]
END
```

FRACTIONP will return **TRUE** if the next two characters are a slash and a digit, respectively. In this case **GET.DENOMINATOR** will remove the slash from **:INPUTSTREAM** and read the number that follows.

The definition of **FRACTIONP** is given by

```
TO FRACTIONP
OP AND SLASHP PEEK.CHAR
       NUMBERP PEEK.2ND.CHAR
END
```

PEEK.2ND.CHAR returns the second character of **:INPUTSTREAM**. If **:INPUTSTREAM** is empty or has only one character in it, then it returns the empty word:

```
TO PEEK.2ND.CHAR
IF EMPTYP :INPUTSTREAM [OP " ]
IF EMPTYP BF :INPUTSTREAM [OP " ]
OP FIRST BUTFIRST :INPUTSTREAM
END
```

Exercise 8.11.1. The slash character, /, is used for two different purposes: to represent the division operation and as a separator between the numerator and denominator of a fraction. Indicate the meaning of each occurrence of / in the following expression assuming the preceding definition of **GET.DENOMINATOR**:

4 /3 + 2/3/5 / 6/7

Exercise 8.11.2. Test **RATSCAN** by

  a. Executing **MAKE "INPUTSTREAM READLINE**.
     Entering **1/2 + 1/3 * 1/4** followed by RETURN.
     Executing **REPEAT 10 [SHOW :INPUTSTREAM**
                               RATSCAN
                               SHOW :TOKEN]

  b. Executing **MAKE "INPUTSTREAM READLINE**.
     Entering **4 /3 + 2/3/5 / 6/7** followed by RETURN.
     Executing **REPEAT 10 [SHOW :INPUTSTREAM**
                                 RATSCAN
                               SHOW :TOKEN]

8.12 The Parser

The parser, named **RATPARSE**, has two inputs: the first token in the expression and a right binding power. It reads everything in the input stream until it encounters an operation with a left binding power that is smaller than that of the right binding power.

RATPARSE is initially called by **RATREAD** with a right binding power of 1. **RATPARSE** will return the representation of the rational-number expression read when it comes across a lower left binding power. Since

298   *Building Your Own Computational Environment*

EOL is the only operation with a left binding power smaller than 1, **RATPARSE** will not return until it reads the entire expression.

Here is the definition of **RATPARSE**:

```
TO RATPARSE :OBJ :RBP
IF RAT.CONSTANTP :OBJ
   [ RATSCAN
     OP RATREADRIGHT :OBJ :TOKEN :RBP]
GEN.RAT.ERROR [NOT A RATIONAL NUMBER:] :OBJ
END
```

RATPARSE is always called at the beginning of an expression. Because an expression always begins with a rational number, an error is generated if something else is present. If **:OBJ** is a rational number, then **RATSCAN** is called to put the operation following the rational number into the global variable **:TOKEN**. We then return the value computed by **RATREADRIGHT**.

RATREADRIGHT takes an expression, the operation following the expression, and a right binding power as inputs. It is responsible for constructing and returning the expression terminated by an operation with a left binding power that is lower than its right binding power input. Here is the definition:

```
TO RATREADRIGHT :EXP :OPER :RBP
LOCAL "TEMP
IF NOT RAT.OPERP :OPER
   [GEN.RAT.ERROR [ILLEGAL OPERATION:] :OPER]
IF (GET.LBP :OPER) < :RBP
   [OP :EXP]
   [RATSCAN
    MAKE "TEMP MK.RATEXP :OPER
                        :EXP
                        RATPARSE :TOKEN GET.RBP :OPER
    OP RATREADRIGHT :TEMP :TOKEN :RBP]
END
```

First, **RATREADRIGHT** checks to make sure it got a legal operation (**+**, **-**, *****, **/**, or **EOL**) from **RATSCAN**. If not, an error is signaled.

If the left binding power of the operation is smaller than the binding power input, it simply returns the expression. This occurs, for example, if **:OPER** happens to be the end-of-line token. That is, we have arrived at the end of the expression.

Otherwise, we need to read the rest of the expression. We find the second argument to **:OPER** (the first argument is **:EXP**) and construct the representation for the expression. This is performed by the sequence of statements

    RATSCAN
    MAKE "TEMP MK.RATEXP :OPER
                            :EXP
                            RATPARSE :TOKEN GET.RBP :OPER

RATSCAN is called to set the value of **:TOKEN** to the first token following **:OPER**. Next, **RATPARSE** is called to read the expression making up the second argument to **:OPER**. This call is given the right binding power of **:OPER** as its second argument to insure it will terminate reading when an operation with a smaller left binding power is encountered. Then **MK.RATEXP** is called to construct and return the representation of the entire expression. Finally, the result is temporarily stored in the local variable named **:TEMP**. For example, suppose that the user entered the expression

    1/2 + 2/3 - 3/4

RATREADRIGHT is called with the following input values:

    :EXP   [1 2]
    :OPER   +
    :RBP   1

The values of **:INPUTSTREAM** and **:TOKEN** are

    :INPUTSTREAM   2/3 - 3/4
    :TOKEN   +

After comparing the left binding power of + with **:RBP**, **RATSCAN** is executed, modifying **:INPUTSTREAM** and **:TOKEN** as follows:

    :INPUTSTREAM   - 3/4
    :TOKEN   [2 3]

300 *Building Your Own Computational Environment*

Next we execute

    RATPARSE [2 3] 6

That will return [2 3], since the - operation has a left binding power smaller than 6. It is also important to note that after the return of this call to **RATPARSE**, the values of **:INPUTSTREAM** and **:TOKEN** are:

    :INPUTSTREAM   3/4
    :TOKEN   -

MK.RATEXP will return the representation of the three parts making up the expression, namely

    [+ [1 2] [2 3]]

After this list is saved in **:TEMP**, we read the rest of the expression looking for an operation with left binding power smaller than **:RBP**. This is performed by calling **RATREADRIGHT** with the list [+ [1 2] [2 3]] as the value of **:EXP**, the value of **:TOKEN** as the value of **:OPER**, and the same **:RBP** as before. That is, the original call to **RATREAD** results in the following simplification:

RATREAD

```
=> RATPARSE [1 2] 1
          with,  :INPUTSTREAM + 2/3 - 3/4
                 :TOKEN        [1 2]

=> RATREADRIGHT [1 2] "+ 1
          with,  :INPUTSTREAM 2/3 - 3/4
                 :TOKEN       +
          RATPARSE [2 3] 6 returns [2 3]
          with,  :INPUTSTREAM 3/4
                 :TOKEN       -
          :TEMP is set to [+ [1 2] [2 3]]

=> RATREADRIGHT [+ [1 2] [2 3]] "- 1
          with,  :INPUTSTREAM 3/4
                 :TOKEN       -
          RATPARSE [3 4] 6 returns [3 4]
          with,  :INPUTSTREAM
```

```
               :TOKEN      EOL
         :TEMP is set to [- [+ [1 2] [2 3]] [3 4]]

=> RATREADRIGHT  [- [+ [1 2] [2 3]] [3 4]]  "EOL  1

=> [- [+ [1 2] [2 3]] [3 4]]
```

† **Exercise 8.12.1.** Define **MK.RATEXP**.

Exercise 8.12.2. Implement the rational-number calculator and test it on the following expressions:

    a. 1/2 + 1/3 + 1/4
    b. 1/2 * 1/3 + 1/4
    c. 1/2 + 1/3 * 1/4
    d. 1/2 + 2 - 1/3
    e. 1/2/3
    f. 1 /2/3
    g. 1/ 2/3
    h. 1/3 + 1/6

8.13 Parenthesized Rational Expressions

Our calculator does not handle grouping of expressions using parentheses. We would like to add this capability to our calculator. The reader is the piece of code that needs to be modified to do this. The scanner then has two additional tokens to look for: a left parenthesis and a right parenthesis, and the parser needs to decide how to handle expressions that include them. First, we need to include them in our BNF description of rational expressions. We do this by modifying the definition of <rat exp> to be

```
        <rat exp> ::=    <constant> |
                         <rat exp> <oper> <rat exp> |
                         <left paren> <rat exp> <right paren>
<oper> ::= + | - | * | /
<constant> ::= <integer> | <integer> <slash> <integer>
```

```
<slash> ::= /
<left paren> ::= (
<right paren> ::= )
```

As tokens, left and right parenthesis are represented by themselves:

```
( ---> (
) ---> )
```

That is, the scanner sets **:TOKEN** to be one of these two characters. The only piece of the scanner that needs modifying is **SCAN**. We need to add two lines

```
TO SCAN :NEXTCHR
IF EMPTYP :NEXTCHAR [OP "EOL ]
IF RAT.OPERP :NEXTCHAR [OP GET.CHAR]
IF LEFT.PARENP :NEXTCHAR [OP GET.CHAR]
IF RIGHT.PARENP :NEXTCHAR [OP GET.CHAR]
IF DIGITP :NEXTCHAR [OP GET.RAT]
IF SPACEP :NEXTCHAR [REMOVE.CHAR OP SCAN PEEK.CHAR]
    [GEN.RAT.ERROR [ILLEGAL CHARACTER IN INPUT:]
                    :NEXTCHAR]
END
```

where **LEFT.PARENP** and **RIGHT.PARENP** return **TRUE** if its input is a left or right parenthesis, respectively, and **FALSE** otherwise.

Passing of parenthesized expressions follows the rules in mathematics. For example,

```
(1/2 + 2/3) * 5  --->  [* [+ [1 2] [2 3]] [5 1]]]
```

The parser looks for the occurrence of a left parenthesis at the appropriate place. Since left parentheses occur where we expect to see a rational number (at the beginning of the expression or following an operation), it is the job of **RATPARSE** to check for it. Here is the new version of **RATPARSE**

```
TO RATPARSE :OBJ :RBP
LOCAL "TEMP
IF RAT.CONSTANTP :OBJ
    [RATSCAN  OP RATREADRIGHT :OBJ :TOKEN :RBP]
IF LEFTPARENP :OBJ
```

```
[ MAKE "TEMP RATPARSEPAREN
  RATSCAN
  OP RATREADRIGHT :TEMP :TOKEN :RBP]
  [GEN.RAT.ERROR [ILLEGAL OBJECT IN EXPRESSION:] :OBJ]
END
```

If a rational number is not recognized at the beginning of the expression, we check for a left parenthesis. If a left parenthesis is discovered, we call **RATPARSEPAREN** to read the parenthesized expression and tuck it away inside the temporary variable, **:TEMP**.

Since the occurrence of the matching right parenthesis should terminate **RATPARSEPAREN**, we assume that the value of **:TOKEN** is the matching right parenthesis after returning from **RATPARSEPAREN**. Thus, we call **RATSCAN** to get the token following the) character. This should be an operation. The value of **:TEMP** is the first argument to this operation, so we call **RATREADRIGHT** to finish reading the expression.

The definition **RATPARSEPAREN** is given by:

```
TO RATPARSEPAREN
RATSCAN
OP RATPARSE :TOKEN 1
END
```

After scanning past the left parenthesis, the value of **:TOKEN** marks the beginning of the parenthesized expression. That is, the value of **:TOKEN** is either a rational number or another left parenthesis. To read the expression beginning with this token we call **RATPARSE** with a binding power of 1. In order to terminate the reading when the matching parenthesis is seen, we treat the right parenthesis token as an operation with a left binding power of 0. That is, a right parenthesis serves the same purpose as **EOL**. Thus, we need to add it to the table of operations. To be consistent with the other operations, we give it a right binding power of 1, even though it will never be used. The operation table should be created as follows:

```
MAKE "OPTABLE [ [+ 5 6] [- 5 6] [* 7 8]
               [/ 7 8] [EOL 0 1] [) 0 1]]
```

Exercise 8.13.1. Define **LEFT.PARENP** and **RIGHT.PARENP**.

Exercise 8.13.2. Add the parenthesized expressions to the rational-number calculator and test it on the following expressions:

    a. 1/2 + 1/3 + 1/4
    b. 1/2 * 1/3 + 1/4
    c. 1/2 + 1/3 * 1/4
    d. 1/2 + 2 - 1/3
    e. 1/2/3
    f. 1 /2/3
    g. 1/ 2/3
    h. 1/3 + 1/6

    i. (1/2 + 1/3) + 1/4
    j. 1/2 * (1/3 + 1/4)
    k. 1/2 + (1/3 * 1/4)
    l. 1/2 + (2 - (1/3 + 1/4))
    m. 1/(2/3)
    n. (1/2 - 1/3)/2*(1/2 + 1/3)
    o. (1/2 + (1/3 / (1/5 - 1)))

Exercise 8.13.3. Suppose we wish to add an operation that allows raising a rational number to an integral power. For example, if we name this operation ^, then the calculator behaves as follows:

RATCALC> 1/2 + 1/2 ^ 2
3/4

How can this be done? What new definitions are needed? What existing procedures need to be redefined?

Chapter 9
Applications And Projects

9.1 Introduction

This chapter outlines several applications and projects. They differ significantly in scope, from symbolic as well as numerical applications in mathematics to general problem solving. They present opportunities to explore and refine the techniques discussed in previous chapters as well as to explore and refine your own problem-solving techniques. They also illustrate the power and flexibility of the Logo programming language.

The exercises and projects are presented in various levels of detail. Those vaguely outlined are done so intentionally. The idea is not to present a collection of complete programs or step-by-step implementation of significant problems. Rather, you should involve yourself in the rigorous definition of the problem at hand as well as in developing your own style of programming.

We have attempted to present a modular construction of programs, isolating the conceptual pieces in their respective program modules. Whether this is done in a top-down, bottom-up, or middle-out fashion will depend on the problem at hand as well as your state of mind. We believe that there is no single programming methodology that works best for all people or all programming applications. Logo allows you the flexibility to develop your own style and freely admits almost any problem-solving strategy.

9.2 A Descriptive Statistics Program

This section discusses the components of a statistical package to evaluate a collection of data. We are concerned with gathering information called **measures of central tendancies** (mean, median, and mode) and **measures of dispersion** (variance and standard deviation) and with constructing graphical displays of the data in frequency diagrams.

We assume that the data collected can be ordered and that a number has been assigned to each item of data. This number, called a **score**, indicates the position of the item relative to other items. For example, if we conduct a survey concerned with rating a computer software product with the extreme responses being poor and excellent, we could assign a score of 0 for poor and 10 for excellent. Other responses (fair, good, very good) could be scored between these values.

Once scores (numbers) have been associated with data items the data can be analyzed by computing numerical quantities based on the scores. The definitions that follow describe the statistical information in which we are interested. The median, mode, and arithmetic mean describe central tendencies, while the range, deviation variance, and standard deviation are measures of dispersion.

> The **median** is a value for which half of the scores collected are above its value and half are below its value.

If the number of scores is an odd number, then the median is the "middle score." If the number of scores is even, the **median** is the number halfway between the two middle scores, or their average.

> The **arithmetic mean** is the average of all the scores.
>
> A **mode** is a score that occurs most frequently.

There may be more than one mode. For example, if in our poll regarding computer programs, we received seven responses of "excellent," seven responses of "very good," and no more than six responses for the other ratings, then there are two modes associated with the data.

The range, the variance, and the standard deviation measure the dispersion of the data. That is, they tell us how the scores are scattered.

> The **range** of the data is the difference between the highest and lowest scores.
>
> The **deviation** of an item of data is defined to be the difference between its score and the **arithmetic mean**.

The deviation of a data item tells us how far a particular response is from the average.

> The **variance** is the sum of the squares of the deviations from the mean divided by the number of data items collected.

The variance gives us a measure of how widely scattered the data are. The deviations are squared, so that they are all positive. To obtain a number that describes how the data are dispersed that is more in line with the scores of the data items, we take the square root of the variance.

> The **standard deviation** is the square root of the variance.

Given a collection of scores, it will soon be apparent that some of the quantities are easier to compute if the scores have been sorted from smallest to largest. Thus, we assume that the data are represented by a list of scores arranged in ascending order. The range, for example, is computed by taking the difference between the last element and the first element.

Project 9.1

Write a Logo program that performs a statistical analysis of a collection of data. The following steps provide a modular approach to the task:

Step 1: For each statistical measure listed, write a Logo function that returns its value. Each function will take at least one input: a list of scores from a collection of data that are ordered from least to greatest.

   a. Range
   b. Median
   c. Arithmetic mean
   d. Modes
   e. Variance
   f. Standard deviation

A useful way to present data is in the form of a **frequency table**. Rather than listing all the scores from least to greatest, we list each score, along with the number of data items for each score.

Step 2: Write a Logo program that takes a list of scores ordered from least to greatest as its only input. It returns a frequency table. The table is represented as a list of lists. Each sublist contains a score and the number of data items that have that score.

When there are 50 or more scores with a large range, the frequency table of scores is usually too scattered to yield a trend. To obtain a more compact view of the data, you can divide the range into several subintervals and compute a frequency table that indicates the number of scores that fall into each subinterval. Such an arrangement is called a **class frequency distribution**.

Step 3: Write a Logo program that takes an ordered list of scores and a number n indicating the number of subintervals of the range to use in computing a class frequency distribution. The program should return a list of lists representing the distribution. Each sublist contains the lowest and highest scores for the subinterval and the number of scores that fall into that subinterval.

Step 4: Define a Logo program that displays the **class frequency distribution** as a histogram. The program takes a class frequency distribution as its only input.

Step 5: Define a Logo program that displays the **class frequency distribution** by connecting the points at the midpoint of the top of each rectangle of the histogram representing the class frequency distribution.

Step 6: Write a Logo program that takes a list of numbers as its only input. It returns a list with the numbers ordered from least to greatest.

9.3 A Calculator for any Number System

The goal of this section is to build a calculator that performs integer addition, subtraction, multiplication, and division, where integers are represented in a notation other than the usual base 10 representation.

The base 10 notation for a number consists of a sequence of characters selected from the collection {0, 1, 2, 3, 4, 5, 6, 7, 8, 9}. The resulting word is called a **numeral**. A numeral is just a name for a number.

A general technique for representing numbers is to choose a finite set of symbols, called an **alphabet**, and order them. The first symbol is used to denote the number 0, the next symbol to denote the number 1, and so on.

For example, we could use a sequence of characters selected from the collection {0 1} to name numbers. The names will become longer than those used with the base 10 representation because we have only two characters in the alphabet.

The numeral formed by a particular string of digits using this notation can be described as follows. The notation, $d_n \ldots d_3 d_2 d_1 d_0$, where $d_0, \ldots, d_n$ are characters from the chosen alphabet is interpreted to mean:

$$d_n * b^n + \ldots + d_2 * b^2 + d_1 * b^1 + d_0 b^0$$

where b is a number called the **base** of the notation. The number base is equal to the number of characters in the alphabet. If b is the number 10, and the alphabet consists of {0, 1, 2, 3, 4, 5, 6, 7, 8, 9 } with the usual ordering, we have *the* base 10 notation for numbers. If b is 2 and the alphabet consists of {0, 1}, we have a base 2 representation of the number.

Base 16 notation requires an alphabet of 16 characters. The convention is to use the ordered collection (0, 1, 2, 3, 4, 5, 6, 7, 8, 9, A, B, C, D, E, F). The digits, A, B, C, D, E, and F denote the numbers 10, 11, 12, 13, 14, and 15.

All computers store numbers using their base 2 representations. It appears that the computer understands base 10 representation, because it accepts the forms that we enter and computes the desired results correctly.

Actually, the computer immediately converts the base 10 representation into a base 2 representation. All computations inside the machine are done using the base 2 representations. If a numeral is to be printed out, the machine quickly computes the sequence of characters, forming the numeral's base 10 representation.

We would like to implement a calculator that will accept arithmetic expressions where numbers are represented in a different notation, compute the result, and display the result in the same notation. For example, here is a sample interaction with a base 2 calculator:

```
CALC2> 101 + 110
1011
CALC2> 1001 - 10 + 101
1100
```

The **TOPLEVEL** procedure for the calculator could be defined as follows:

```
TO CALC2
TYPE "CALC2\>
CALC2PRINT CALC2RUN CALC2READ
CALC2
END
```

After the prompt **CALC2>** is displayed, **CALC2READ** is called to read the expression entered by the user and represent it as a Logo expression that can be evaluated by **CALC2RUN**.

Since Logo expects the names of numbers to be written in their base 10 notation, **CALC2READ** has to convert all numerals from their base 2 form to their base 10 equivalent before the expression is simplified by **CALC2RUN**.

A Calculator for any Number System   311

After the value has been obtained, the result is displayed by **CALC2PRINT**. Again, Logo returns the base 10 representation of the number, so **CALC2PRINT** has to convert the base 10 representation into its base 2 representation.

We have outlined the work involved in finishing the calculator. One of the tasks is to convert one representation of a number into another representation. We next develop the algorithms that perform these conversions.

Suppose that we want to convert the number 85's base 10 representation into its base 2 equivalent. This means that we want to find the digits d0, d1, ..., dn, where the d_i come from the set {0, 1}, such that

$$d_n * 2^n + ... + d_2 * 2^2 + d_1 * 2^1 + d_0 * 2^0 = 85$$

If we divide both sides by 2 we see that d_0 is the remainder of 85 divided by 2. We denote this quantity by remainder (85,2). Thus, we have

d_0 = remainder (85,2)

$$d_{n-1} * 2^{n-1} + ... + d_2 * 2^1 + d_1 * 2^0 = \text{quotient} (85,2)$$

Now the base 2 representation of 85 can be described as the base 2 representation of **QUOTIENT (85, 2)** appended to **REMAINDER (85, 2)**. The algorithm can be stated as follows:

Algorithm   To convert the base 10 representation of a number to the base 2 representation of that number, perform the following operations:

1. If the base 10 representation of the number is 0, then return 0.

2. Otherwise, append the base 2 representation of the integer quotient of the number and 2, onto the front of the base 2 representation of the remainder.

Example 9.3.1. Define a Logo function named **CONVERT.10.TO.2**, which takes the base 10 representation of a number as its input and returns the base 2 representation of the number. For example,

CONVERT.10.TO.2 19 returns 10011

The definition of **CONVERT.10.TO.2** is:

```
TO  CONVERT.10.TO.2 :NUM
IF EQUALP :NUM 0 [OP "0]
    [OP WORD CONVERT.10.TO.2 QUOTIENT :NUM 2
            REMAINDER :NUM 2]
END
```

The definition has one minor flaw, as can be seen by tracing the evaluation of **CONVERT.10.TO.2 19**.

A zero will always be appended onto the front of the representation. Mathematically speaking, the zero does not matter, but the result is undesirable. We can fix the problem with the following definition:

```
TO  BASE.10.TO.2 :NUM
IF EQUALP :NUM 0 [OP "0] [OP CONVERT.10.TO.2 :NUM]
END
```

where we modify **CONVERT.10.TO.2** as follows:

```
TO  CONVERT.10.TO.2 :NUM
IF EQUALP :NUM 0 [OP " ]
    [OP WORD CONVERT.10.TO.2 QUOTIENT :NUM 2
            REMAINDER :NUM 2]
END
```

This technique works for converting the representation of a number in one base to the representation of the number in any other desired base. This general algorithm can be stated as follows.

Algorithm   To convert the base b representation of a number to the base c representation of that number, do the following:

1. If the base b representation is 0, then return 0.

2. Otherwise, append the base c representation of the integer quotient of the number with c onto the front of the base c representation of the remainder.

Exercise 9.3.1. Define a Logo function, named **BASE.10.TO.8** that requires the base 10 representaion of a number as its input and returns the base 8 representation of the number. For example,

    BASE.10.TO.8 19 returns **23**

Exercise 9.3.2. Define a function named **BASE.10.TO.16** that takes the base 10 representation of a number as its input and returns the base 16 representation. Recall that a base sixteen number requires the use of the digits 0, 1, 2, 3, 4, 5, 6, 7, 8, 9, and the letters A, B, C, D, E, and F, where A through F represent the numbers 10 through 15.

($Hint$: **REMAINDER :NUM 16** returns the base 10 representation of a number between 0 and 15. If n is the result, then the base 16 character representing this number is the nth character of the base 16 alphabet.)

† **Exercise 9.3.3.** Define a function named **BASE.10.TO.C** that takes two inputs: the base 10 representation of a number and an integer representing the desired base (2 through 16) for representing the number. It returns the representation of the number in the new base. For example,

    BASE.10.TO.C "19 "8

requests the base 8 representation of the number 19. It returns **23**.

† **Exercise 9.3.4.** Define a function named **BASE.C.TO.10** that takes two inputs: the base c ($2 \le c \le 16$) representation of a number and the base used for the representation. It returns the base 10 representation of the same number. For example,

    BASE.C.TO.10 "23 "8

requests the base 10 representation of the number represented by the base 8 notation 23. It returns **19**.

Exercise 9.3.5. Define a function named **BASE.B.TO.C** that takes three inputs: the representation of a number, the base in which the number is given, and the base for the new representation. It returns the representation of the number in the new base. For example,

    BASE.B.TO.C "2B "16 "10

requests the base 10 representation of the base 16 numeral **2B**. It returns **43**. Both bases are assumed to be no greater than 16.

Project 9.2
Finish writing the program for the base 2 calculator described at the beginning of this section in three steps:

    † **Step 1:** Assume that the operations recognized by the calculator are addition, subtraction, and multiplication and the user will never enter bogus base 2 numerals like 317.

    Step 2: Modify your program so that bogus numerals such as 317 result in an "illegal number notation" error without leaving the base 2 calculator.

    Step 3: Modify your program so that it performs integer division.

Project 9.3
Write a program named **CALC.ANY**, which implements a base b calculator for any base from 2 through 16. It takes a number indicating the particular base to be used as an input. For example, executing

    ?CALC.ANY 16

results in a base 16 calculator,

    CALC16>

whereas

    ?CALC.ANY 8

results in a base 8 calculator,

    CALC8>

Displaying the prompt, reading expressions, and displaying results now take the number base as inputs. The toplevel procedure is a slight modification of **CALC2**:

```
TO CALC.ANY :BASE
TYPE MK.PROMPT :BASE
CALC.ANY.PRINT RUN CALC.ANY.READ :BASE :BASE
CALC.ANY :BASE
END
```

Note that the number base being used is an input to **CALC.ANY.READ** and **CALC.ANY.PRINT**.

9.4 Solving Equations

In this section we develop a general technique for solving an equation. That is, given a function $f(x)$, we find all values of x for which $f(x) = 0$. We start with a more specific problem:

Example 9.4.1 Define a function named **SQRT1**,[*] which returns the square root of a positive number.

Finding the square root of a number cannot be done by forming a simple combination of the primitve operations provided by Logo. We must develop a program that computes a close approximation.

The problem is equivalent to solving the equation

$$x^2 = a$$

That is, the square root of the positive number a is found by finding the values of x such that $x^2 = a$. We assume that we are looking for the positive value of x that satisfies this equation. If we define the function $f(x) = x^2 - a$, we can reformulate the problem as solving the equation $f(x) = 0$.

---

[*] We use the name **SQRT1** to distinguish it from the Logo primitive named **SQRT**.

If we examine the graph of f(x) we can see that it appears that the square root of a is trapped between 0 and a. That is, f(0) is negative, whereas f(a) is positive. Furthermore, we can determine whether or not $\sqrt{a}$ lies in the interval [0, a / 2] or [a / 2, a] by examining the sign of f(a / 2). If f(a / 2) > 0, then $\sqrt{a}$ lies in [0, a / 2]. Otherwise $\sqrt{a}$ lies in [a / 2, a].

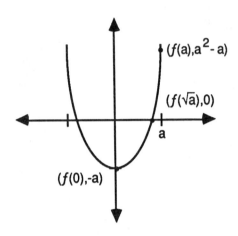

We can continue the process, isolating $\sqrt{a}$ in a smaller and smaller interval. We cannot carry out this process indefinitely, but we can carry out the process until we have trapped $\sqrt{a}$ in an interval as small as we like, thereby obtaining an approximation to $\sqrt{a}$ as accurately as desired.

We have developed a more general algorithm than we were looking for. That is, the scheme can be used to find the solution of many problems that involve finding where the value of a function is zero.

Suppose we have a function f defined on an interval $[x_l, x_r]$ that satisfies the following properties:

1. $f(x_l) * f(x_r) < 0$

2. $f(x)$ is continuous on $[x_l, x_r]$. That is if the graph is a continuous curve with no gaps, then we can find an approximation to the solution of $f(x) = 0$ as accurately as we like by using the following algorithm.

Algorithm (Successive Halving). If $f(x)$ satisfies the given conditions and we desire to find an approximation to the solution of $f(x)$ with an error that is no greater than e, the following function returns the desired approximation, where x_m denotes the midpoint of the interval $[x_l, x_r]$:

$$solve(f, x_l, x_r, e) = \begin{cases} x_m \text{ if } abs(x_r - x_l) < e \\ solve(f, x_l, x_m, e) \text{ if } f(x_l)*f(x_m) < 0 \\ solve(f, x_m, x_r, e) \text{ if } f(x_l)*f(x_m) > 0 \end{cases}$$

We can apply this algorithm to the square root problem. All we need to do is call solve with $x^2 - a$ as the function input, 0 as the left endpoint, a as the right input, and whatever we like for the accuracy. That is,

$sqrt(a) = solve(x^2 - a, 0, a, .00001)$

We have made one serious oversight. If a is in the interval $[0, 1]$, $\sqrt{a}$ does not lie between 0 and a. For example, $\sqrt{1/4}$ is 1/2. However, if a is in $(0, 1)$, we do know that $\sqrt{a}$ lies between 0 and 1. We can modify our algorithm by redefining **sqrt** as follows:

$$sqrt(a) = \begin{cases} solve(x^2 - a, 0, a, .00001) \text{ if } a \geq 1 \\ solve(x^2 - a, 0, 1, .00001) \text{ if } 0 < a < 1 \end{cases}$$

The Logo definition of *sqrt* is as follows:

```
TO SQRT1 :A
IF :A = 0 [OP 0]
IF :A < 1 [OP SOLVE [X * X - A] 0 1 .00001]
          [OP SOLVE [X * X - A] 0 :A .00001]
END
```

where **SOLVE** is given by

```
TO SOLVE :F :XL :XR :E
  LOCAL "XM
  MAKE "XM (:XL + :XR)/2
  IF ABS (:XR - :XL) < :E [OP :XM]
  IF (EVALFUN :F :XL) * (EVALFUN :F :XM) < 0
    [OP SOLVE :F :XL :XM :E]
    [OP SOLVE :F :XM :XR :E]
END
```

Note that we have used the **EVALFUN** function defined in Chapter 4 to evaluate the function for a particular value of x. It was developed in the graphing application program.

Exercise 9.4.1. Test the successive halving method in locating solutions to $f(x) = 0$ for each of the following functions:

a. $x^2 - 2$ c. $x^2 - 2x - 15$

b. $x^3 - 3$ d. $\sin x$

Project 9.4

Combine the **EXPLORE.GRAPH** program described in Chapter 7 and the successive halving program into an interactive program that allows the user to explore the graph of $f(x) = 0$ to locate an interval $[x_l, x_r]$ in which the function changes sign and then allows him or her to invoke **SOLVE** to find the solution more accurately.

The program should provide instructions on how to use the program, information about the commands available at any time during the running of the program and should prompt the user for the needed information.

For example, when the program starts up, it should inform the user that he or she can either explore the graph of some function or go directly to the equation solver to search for solutions. While exploring the graph of a function, the program should inform the user of the available commands. Try out the program on the functions defined in Exercise 9.4.1.

9.5 A Set Calculator

In Chapter 6 we defined Logo programs that performed set operations (for example, union, intersection, and set difference), where sets were represented as lists. In this section we outline a project that implements a set calculator. The set calculator prompts the user for an expression, evaluates it, and displays the result. That is, we want to write a program that reads in set expressions where the value is displayed in set notation, rather than as lists. A typical interaction might look like this:

    SETCALC> {A B C} ∪ {A D E}
    {A B C D E}

    SETCALC> {A B} ∩ {B C D} ∪ {C D}
    {B}

    SETCALC> {A B C D E} - {B D}
    {A C E}

    SETCALC>

where **SETCALC>** is the set calculator's prompt, ∪ denotes set union, ∩ denotes set intersection, and - denotes set difference.

To write the **SET CALCULATOR** we need to define a **TOPLEVEL** loop that handles the prompting, reading, and printing:

    TO SET.TOPLEVEL
    TYPE "SETCALC\>
    SET.PRINT SET.RUN SET.READ
    SET.TOPLEVEL
    END

SET.READ reads the expression typed in by the user. It returns a representation of the set expression that can be simplified by **SET.RUN**. The result returned by **SET.RUN** is the representation of a set, which is displayed by **SET.PRINT**. To implement the set calculator, we need to define **SET.PRINT**, **SET.READ**, and **SET.RUN**.

Exercise 9.5.1. Implement the set calculator by defining **SET.PRINT**, **SET.READ**, and **SET.RUN** as follows:

> a. **SET.PRINT** takes the representation of a set and displays it in set notation. In Chapter 6, we chose to represent sets by lists, so **SET.PRINT** simply displays a left brace, or curly bracket, the elements of the set, and the closing brace. Remember that sets may occur as elements of a set.
>
> b. **SET.READ** consists of a scanner named **SET.SCAN** and a parser named **SET.PARSE**. The scanner reads the characters in the input stream and return a token of a set expression. **SET.PARSE** calls **SET.SCAN** to get the tokens and build the representation of a set expression.

Before we write **SET.READ**, we specify the pieces more carefully. The expressions the reader expects to find are described as follows:

<set exp> ::= <set> | <term> <set.op> <term>
<term> ::= <set exp>
<set.op> ::= ∪ | ∩ | -

This says that a set expression is either a set or a combination of sets that contain any number of set operations and that we limit ourselves to three set operations: union, intersection, and set difference. Some examples are:

{1 2 3}
{1 2 3} ∪ {2 4 5 3 8}
{1 2 3} ∩ {2 4 5 3 8}
{1 2 3} - {2 4 5 3 8} ∪ {6 7 8}
{1 2 3} ∪ {7 8} ∩ {3 5 7}

There are five tokens with which we need to deal: sets, the symbols ∪, ∩, and -, and the word **ENDEXP**, which is returned by the scanner when the input has been exhausted.

The reader reads this text and returns a form that **SET.RUN** can handle. Since we have already defined Logo functions to perform union, intersection, and set difference, the reader returns the following lists for each of the given examples, respectively:

[1 2 3]
[UNION [1 2 3] [2 4 5 3 8]]
[INTERSECTION [1 2 3] [2 4 5 3 8]]
[UNION SETDIFF [1 2 3] [2 4 5 3 8] [6 7 8]]

[UNION [1 2 3] INTERSECTION [7 8] [3 5 7]]

That is, the representation for set expressions can be described as follows:

<set> -----> <list>
<term> <set op> <term> -----> [<set op> <term> <term>]

For example, **SET.READ** reads an expression like

{A B C} ∪ {B D E}

and returns the list

[UNION [A B C] [B D E]]

The reader also assumes that union has binding powers of 5 and 6, that intersection has binding powers of 7 and 8, and that set difference has binding powers of 5 and 6, and in a left-to-right order when binding powers are the same.

   c. **SET.RUN** takes the representation of a set expression returned by **SET.READ** and returns the simplified result. That is, **SET.RUN** returns a list representing a set.

Exercise 9.5.2. Modify the set calculator to include parenthesized set expressions.

Exercise 9.5.3. Modify the set calculator so that set complement with respect to the letters of the alphabet is an allowable operation. Use * to denote set complement. It is used as a postfix operation. For example the complement of {BCD} is shown by,

{B D E}*

Exercise 9.5.4. Modify the set calculator so that it will catch and handle all errors generated by the evaluator. That is, it will not abort and return to Logo's top level.

9.6 Missionaries and Cannibals

The missionaries and cannibals problem begins with a group of three missionaries and three cannibals on the jungle side of a river. The problem is to get all six people to the city side of the river without losing any missionaries. There is a boat on the jungle side of the river, which can carry at most two people for a single trip. The boat must have a pilot as well. Thus, at least one person must be on the boat when it crosses the river.

As long as the number of missionaries on one side of the river is at least equal to the number of cannibals, the cannibals do not attempt to do what comes naturally. Can we get all six people safely across the river? If so, what sequence of moves leads to the solution?

The answers to these questions can be obtained by a process of elimination. We begin by performing all possible first moves. Any resulting state that is unstable is eliminated. From each of the resulting safe states, we perform all possible moves, again eliminating those that are not safe. We perform these actions until we stumble on the desired state or have exhausted all the possibilities.

This strategy can be cast in the form of a tree of changing states. The initial state has all six people and the boat on the jungle side of the river. The set of possible next moves (not necessarily safe) can be represented as the branches from the initial state to the states arrived at by performing each of those moves. Since there are five possible first moves, we have the tree pictured in Figure 9.2.

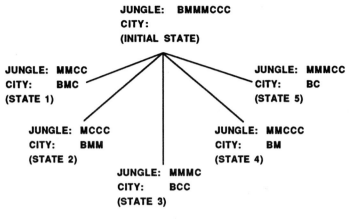

Figure 9.2

Since it is clear that two of the possible first moves (state 2 and state 4) will result in disaster, only moves from each of the possible safe states will be examined. If all paths lead to the eventual loss of a missionary, we will discover that the problem has no successful conclusion. If we find ourselves at our goal state (all six people on the city side of the river), we will have discovered the problem has a solution.

There is one pitfall to this strategy. We could easily perform a sequence of moves that simply alternates between two safe states. That is, we may be going around in circles forever. For example, the move that leads us to state 5 in Figure 9.2, while safe, has only one possible next move: the boat and the cannibal must return to the jungle side of the river. But this leaves us with the original state. Performing these moves over and over again is safe, but it does not get us anywhere. We must modify our strategy by remembering all states that have been seen so far. If we arrive at a state we have seen before, we eliminate it as we would a bad move.

Our first concern is whether or not a solution exists. Therefore, we formulate an algorithm that simply tells us whether or not a path leading to the goal state exists.

Our search begins with an initial state. We replace that state by the five states at the next level of the state tree. Thus, the algorithm has a collection of states to examine. It proceeds as follows:

1. If the first state is the goal state, return **TRUE**.

2. If the first state is an unstable one, throw it away and examine the remaining states.

3. If we have seen the first state before, examine the remaining states.

4. If we have not seen the first state before and it is stable, remember it as one we have seen and replace it with the collection of states obtained by performing all possible moves from it.

5. If we have examined all possible paths in the tree without having reached our goal state, then the problem has no solution and **FALSE** is returned.

The Logo program to determine whether or not a solution exists is named **SOLVE.MCP**. It has a list of states to examine as its only input. Initially, the list contains the initial state. We add new states to examine and dispense with those that do not lead us anywhere. Thus, we have failed when this list becomes empty. Here is the definition of **SOLVE.MCP**:

```
TO SOLVE.MCP :STATELIST
LOCAL "TEST.STATE
IF EMPTYP :STATELIST [OP "FALSE]
MAKE "TEST.STATE FIRST :STATELIST
IF REACHED.GOAL :TEST.STATE [OP "TRUE]
IF UNSTABLE :TEST.STATE [OP SOLVE.MCP BF :STATELIST]
IF SEEN.BEFORE :TEST.STATE
   [OP SOLVE.MCP BF :STATELIST]
   [ REMEMBER :TEST.STATE
     OP SOLVE.MCP REPLACE.FIRST.STATE :STATELIST]
END
```

Since the first state in the list is used in several places, we use a local variable, **TEST.STATE**, to hold the value of the state being examined. The definitions of **REACHED.GOAL** and **UNSTABLE** are:

```
TO REACHED.GOAL :STATE
OP 6 = (MISSIONARIES.ON CITYSIDE :STATE) +
       (CANNIBALS.ON CITYSIDE :STATE)
END
```

```
TO UNSTABLE :STATE
OP OR MISSIONARYLOST CITYSIDE :STATE
      MISSIONARYLOST JUNGLESIDE :STATE
END
```

The **CITYSIDE** and **JUNGLESIDE** functions take a state as input and return the state on the city side and jungle side, respectively. **MISSIONARIES.ON** takes the state on one side of the river and returns the number of missionaries on that side. **CANNIBALS.ON** takes the state on one side of the river and returns the number of cannibals on that side. The **MISSIONARYLOST** function returns **TRUE** if there are more cannibals than missionaries on one side of the river and **FALSE** otherwise.

The definitions of the functions used in **REACHED.GOAL** and **UNSTABLE** depend on the representation for a state. These details are left at a lower level of coding. The only functions that care about what a state looks like are: **MISSIONARIES.ON**, **CANNIBALS.ON**, **CITYSIDE**, and **JUNGLESIDE** (see Exercise 9.6.1).

SEEN.BEFORE checks to see if we have already encountered this state in our search. Its definition will depend on how we **REMEMBER** previously seen states (see Exercise 9.5.2).

The definition of **REPLACE.FIRST.STATE** is as follows:

```
TO REPLACE.FIRST.STATE :STATELIST
OP SE MK.NEW.STATES FIRST :STATELIST
      BF :STATELIST
END
```

This function takes the current state list as its input and returns a new one constructed by replacing the first with the states that are reachable from it in one move. **MK.NEW.STATES** performs this task:

```
TO MK.NEW.STATES :STATE
OP MK.NEW.STATE.LIST :STATE GENMOVES :STATE
END

TO MK.NEW.STATE.LIST :STATE :MOVELIST
IF EMPTYP :MOVELIST [OP [ ]]
   [OP FPUT MK.STATE FIRST :MOVES :STATE
             MK.NEW.STATE.LIST :STATE BF :MOVELIST]
END
```

The **GENMOVES** function takes a state as an input and generates a list of possible next moves from the input state. **MK.NEW.STATE** takes a move and a state as inputs and returns the state representing the completion of the move.

Exercise 9.6.1. We can represent a state as a two-element list. The first element is a list denoting the state on the jungle side of the river and the second element is a list denoting the state on the city side of the river. Each list contains an **M** for each missionary on that side of the river, a **C** for each cannibal, and a **B** if the boat is on that side. For example, the list

[[M M C] [B M C C]]

indicates that two missionaries and a cannibal are on the jungle side of the river and the boat, one missionary, and two cannibals are on the city side of the river.

Define the following functions using this representation:

a. **CITYSIDE** *state*
b. **JUNGLESIDE** *state*
c. **MISSIONARIES.ON** *stateofonesideoftheriver*
d. **CANNIBALS.ON** *stateofonesideoftheriver*
e. **MISSIONARYLOST** *stateofonesideoftheriver*

Exercise 9.6.2. Define **REMEMBER** and **SEEN.BEFORE** for each of the following assumptions:

   a. States seen before are stored using a global variable named **REMEMBERED.STATES**.

   b. States seen before are stored on the property list of the word **STATE** under the **SEEN** property.

Exercise 9.6.3. The functions **GENMOVES** and **MK.STATE** depend on the representation for a move. Describe a representation for a move and define them. The descriptions of **GENMOVES** and **MK.STATE** are:

   GENMOVES takes a state as its input and returns a list of possible moves from that state.

MK.STATE takes a move and a state as inputs and returns the state obtained by performing the move from the given state.

Exercise 9.6.4. Determine if the missionaries and cannibals problem has a solution by implementing **SOLVE.MCP** and executing it.

If you have run the program that tests for a solution to the missionaries and cannibals problem and found that it has a solution, you are now ready to compute the sequence of moves that solves the problem. An equivalent task is to construct the sequence of states leading to the goal.

The **SOLVE.MCP** program keeps track only of the states to be examined next. It does not keep track of the paths traveled in the state tree. The solution to finding the path is just a slight modification of **SOLVE.MCP**. We want the inputs to the program to be the paths being traveled rather than the most recent states of each path.

A path can be represented as a list of states, with the most recent state being the first element of the list. The program to return the path leading to a solution is named **SOLVE.MC**. It takes a list of paths taken so far as its input. Its definition is as follows:

```
TO  SOLVE.MC  :PATHLIST
LOCAL  "TEST.PATH
IF  EMPTYP  :PATHLIST  [OP  [NO  SOLUTION]]
MAKE  "TEST.PATH  FIRST  :PATHLIST
IF  REACHED.GOAL  :TEST.PATH  [OP  TEST.PATH]
IF  UNSTABLE  :TEST.PATH  [OP  SOLVE.MC  BF  :PATHLIST]
IF  SEEN.BEFORE  :TEST.PATH  [OP  SOLVE.MC  BF  :PATHLIST]
   [REMEMBER  :TEST.PATH
    OP  SOLVE.MCP  REPLACE.FIRST.PATH  :PATHLIST]
END
```

The algorithm is the same as for **SOLVEMCP**, except that the algorithm returns a list rather than **TRUE** or **FALSE**. If all paths have been exhausted, a list containing the words **NO SOLUTION** is returned.

REACHED.GOAL checks the most recent state in the first path to determine if the test path has yielded a solution.

UNSTABLE checks the instability of the most recent state in the test path.

SEEN.BEFORE checks to see if the most recent state in the test path has been previously recorded.

REMEMBER saves away the last state in the test path.

Finally, **REPLACE.FIRST** replaces the first path in the path list by paths from the initial state to the states obtained by performing all possible moves from the most recent state in the test path.

Project 9.6.
Which functions in the **SOLVE.MCP** program need to be modified in order to implement **SOLVE.MC**? Redefine them so that **SOLVE.MC** produces the path describing the solution to the missionaries and cannibals problem.

Project 9.7.
Write a program that graphically displays the sequence of moves that lead to a solution. You will need a graphical representation for the boat and for each of the missionaries and cannibals. Your program should show the group moving across the river.

9.7 Newton's Method

Another algorithm for finding the zeros of a function has its derivation in calculus. The solution to the equation $f(x) = 0$ can be approximated by computing a sequence of points $x_0, x_1, x_2, ...,$ where x_0 is any initial approximation to the solution and the remaining points in the sequence are computed by

$$x_n = x_{n-1} - f(x_{n-1})/f'(x_{n-1}) \text{ for } n = 1, 2, ...$$

The sequence of points so generated converges (get closer and closer) to the solution $x*$ provided that $f(x)$ is differentiable, $f'(x*) \neq 0$, and x_0 is "close enough" to $x*$. This technique for finding the solution to $f(x) = 0$ is called **Newton's method**. For many functions, close enough means any guess at all. This is the case for the square root function. We may obtain an approximation to the solution with a prescribed accuracy by computing successive points in the sequence until two successive points become sufficiently close to each other. We may describe the algorithm by:

newton(f, x_0, e) = newton1(f, x_0, nextx(f, x_0), e)

where *nextx* computes the next point in the sequence and *newton1* is defined by

$$newton1(f, x_n, x_{n1}, e) = \begin{cases} x_{n1}, & \text{if } abs(x_{n1}\text{-}x_n) < e \\ newton1(f, x_{n1}, nextx(f, x_{n1}), e), & \text{otherwise.} \end{cases}$$

Exercise 9.7.1. Write a program that approximates the derivative of a function at a point using the following approximation:
f(x+.01) - f(x) /.01.
The function should be input as a list and evaluated using **EVALFUN** (see Section 4.10 for the representation of functions and the definition of **EVALFUN**).

† **Exercise 9.7.2.** Write a Logo program that uses Newton's method to find the solution to an equation of the form $f(x) = 0$. Test it by trying to locate the solutions to $f(x) = 0$ for each of the following functions:

a. $f(x) = x^2 - 2$ c. $f(x) = x^2 - 2x - 15$

b. $f(x) = x^3 - 3$ d. $f(x) = \sin x$

Exercise 9.7.3. Write a Logo program that uses Newton's method to find the cube root of a number.

Exercise 9.7.4. Compare the number of iterations to get an accuracy of 0.0001 first using successive halving and then using Newton's method in computing the square root of a number.

9.8 Symbolic Differentiation

The derivative of a function is another function. Getting numerical approximations to the derivative of a function at a point is a simple problem for "nice" functions. By nice, we mean continuously differentiable functions like polynomials and the trigonometric functions. Obtaining the function itself can be done as well, provided we can conveniently represent such functions. In this section, we develop a program for computing the derivative of a function. That is, the program takes the representation of a function as its input and returns the representation of its derivative.

The following list summarizes the rules for differentiation. The notation D_x is used to denote the derivative of the parenthesized expression with respect to the variable x.

1. Constant rule

    $D_x(c) = 0$

2. Variable rule

    $D_x(v) = \begin{cases} 1 & \text{if } v = x \\ 0 & \text{if } v \neq x \end{cases}$

3. Sum rule

    $D_x(f + g) = D_x(f) + D_x(g)$

4. Product rule

    $D_x(f \bullet g) = f \bullet D_x(g) + g \bullet D_x(f)$

5. Quotient rule

    $D_x(f / g) = (D_x(f) \bullet g - f \bullet D_x(g)) / g^2$

6. Power rule

    $D_x(x^n) = n \bullet x^{n-1}$

7. Chain rule

$$D_x(f^n) = n \cdot f^{n-1} \cdot D_x(f)$$

Note that these rules are recursive. For example, "the derivative of the sum of two functions is the sum of the derivatives of each." The definition of a Logo function to compute the derivative is a straightforward translation of these rules. Here is the definition that incorporates the first four rules assuming the we have recognizers (**IS.CONSTANT, IS.VARIABLE, IS.SUM,** and **IS.PRODUCT**) that tell us which rules to use, selectors (**ARG1,** and **ARG2**) that allow us to select the components of a sum or product, and constructors (**MK.SUM, MK.PROD**) that return the representation of the sum or product of two functions:

```
TO DIFF :FUN :VAR
IF IS.CONSTANT :FUN [OP 0]
IF AND IS.VARIABLE :FUN :FUN = :VAR [OP 1]
IF IS.VARIABLE :FUN [OP 0]
IF IS.SUM :FUN [OP MK.SUM  DIFF ARG1 :FUN :VAR
                           DIFF ARG2 :FUN :VAR]
IF IS.PRODUCT :FUN
    [OP MK.SUM MK.PROD ARG1 :FUN
                       DIFF ARG2 :FUN :VAR
               MK.PROD ARG2 :FUN
                       DIFF ARG1 :FUN :VAR
END
```

To complete this program, we need to decide on a representation for functions such as polynomials and the trigonometric functions as well as define the selectors and recognizers that depend on the representation. There are several ways this can be done, and we choose the following method because it gives us the generality we want:

1. Constant functions are represented by the constant that is its definition. For example, the function, $f(x) = 3$ is represented by the number **3**.

2. Functions that are defined as a variable are represented as that variable. For example, the functions $f(x) = x$ and $g(x) = y$, are represented by the words **X** and **Y**, respectively.

3. The sum of two functions will be represented as a list of three elements. The first element is the word **PLUS**, and the second and third elements are representations of the functions that are being added. For example, the function $f(x) = x + 3$ is represented by the list **[PLUS X 3]**.

4. The product of two functions is represented as a list of three elements. The first element is the word **TIMES**, and the second and third elements are the representations of the functions being multiplied. For example, the function $f(x) = 2x$ is represented by the list **[TIMES 2 X]**.

This representation allows us to represent any polynomial. For example, the polynomial

$2x^2 + 3x$

can be represented as

[PLUS [TIMES 2 [TIMES X X]] [TIMES 3 X]]

The definitions of the selectors, constructors, and recognizers can now be given. For example, the constructor **MK.SUM** takes the two summands as inputs and constructs a list whose elements are the word **PLUS** and the two summands:

```
TO MK.SUM :SUMMAND1 :SUMMAND2
OP (LIST "PLUS :SUMMAND1 :SUMMAND2)
END
```

We can modify **DIFF** so that it handles the other rules as well. For example, to include differentiation rule 6, we would add the following line:

```
IF IS.POWER :FUN
   [OP MK.PROD GET.EXPONENT :FUN
               MK.POWER GET.VARIABLE :FUN
                        (GET.EXPONENT :FUN) - 1]
```

† **Exercise 9.8.1.** Define the recognizers: **IS.CONSTANT, IS.VARIABLE, IS.SUM, IS.PRODUCT,** and **IS.POWER**.

† **Exercise 9.8.2.** Define the selectors: **ARG1, ARG2, GET.EXPONENT,** and **GET.VARIABLE**.

Exercise 9.8.3. Define the constructors: **MK.PROD** and **MK.POWER**.

Exercise 9.8.4. Modify **DIFF** so that it includes differentiation rules 5 and 7.

Exercise 9.8.5. Modify **DIFF** so that it includes the following differentiation rules:

$$D_x (\sin f) = \cos f \bullet D_x (f)$$
$$D_x (\cos f) = - \sin f \bullet D_x (f)$$

Exercise 9.8.6. Test **DIFF** on the following functions:

a. $x^3 + 5x + 3$
b. $(x^3 + 5x + 3)^7$
c. $\sin(x^3 + 5x + 3) + \cos^2 2x$
d. $x / (x-2)$

Exercise 9.8.7. **DIFF** correctly differentiates a large number of functions, but the result is unsimplified. For example, terms and factors such as $0 \cdot X$, $1 \cdot 4$, $7 + 0$, and X^0 appear in the result. These should be simplified to 0, 4, 7, and 1, respectively.

Because the representation of a form like **7 + 0** is constructed in only one place, namely, inside **MK.SUM**, and the representations of forms **0 · X** and **1 · 4** are constructed by **MK.PROD**, we can prevent these forms from occurring in the result by modifying the constructors so that they return the simplified form.

For example, we want **MK.SUM** to return the number **7** if its inputs are **0** and **7**. In general, if both inputs are numbers, we want **MK.SUM** to return the sum of the numbers rather than a list. We can modify **MK.SUM** to handle these simplifications as follows:

```
TO MK.SUM :SUMMAND1 :SUMMAND2
IF :SUMMAND1 = 0 [OP :SUMMAND2]
IF :SUMMAND2 = 0 [OP :SUMMAND1]
IF AND NUMBERP :SUMMAND1
       NUMBERP :SUMMAND2
   [OP :SUMMAND1 + :SUMMAND2]
   [OP (LIST "PLUS :SUMMAND1 :SUMMAND2)]
END
```

a. Modify **MK.PROD** so that it returns a simplified result when at least one of its inputs is **0**, when at least one of its inputs is **1**, or when both inputs are numbers.

b. Modify **MK.POWER** so that it returns a simplified result when the exponent input is **0** or **1**.

c. Test the modifications in **MK.SUM**, **MK.PROD**, and **MK.POWER** on the examples in Exercise 9.8.6.

Appendix A
Some Theory of Computation

A.1 The Halting Problem

By now you ought to be impressed by the power of computing. There are, however, limitations.

In these first two sections, we restrict our attention to programs that compute functions of one input. The simplest question we can ask of a function is if it is always defined, i.e., if it always halts and outputs a value on any given (or any fixed or any appropriate) input. Not every function has this property. For example, consider

```
TO CYCLE.FOREVER :X
OP CYCLE.FOREVER :X - 1
END.
```

This program, not having an escape clause, cycles forever on numerical input and halts without output on nonnumerical input. This was a fairly easy example; they are not always so simple.

> The **halting problem** is the problem of determining whether or not a given program halts on given input.

The halting problem happens to be an unsolvable one: there is no general program that, fed the text of a program as input, outputs **TRUE** if the input program on specified input (say, **0**) will halt and **FALSE** otherwise. This is a fairly simple thing to prove but not so simple a thing to understand.

Theorem The halting problem is unsolvable.

Proof: Suppose we had a program **HALT.CHECKER**, which took the texts of programs as input and output **TRUE** if the given program would halt on input **0** and **FALSE** otherwise. Define a new program **ANTI.HALT.CHECKER** as follows:

```
TO ANTI.HALT.CHECKER :X
IF HALT.CHECKER TEXT "ANTI.HALT.CHECKER
   [OP CYCLE.FOREVER 0] [OP 1]
END.
```

Observe that if **ANTI.HALT.CHECKER 0** halts, then **HALT.CHECKER** says that it halts and, by definition **ANTI.HALT.CHECKER 0** calls and runs **CYCLE.FOREVER 0**, whence it never halts. On the other hand, if **ANTI.HALT.CHECKER 0** never halts, **HALT.CHECKER** tells us so and, by definition, **ANTI.HALT.CHECKER** immediately outputs **1**.

Thus, if a **HALT.CHECKER** program existed, we could define a program **ANTI.HALT.CHECKER** that would halt if and only if **HALT.CHECKER** didn't halt. As this last statement is self-contradictory, we must reject the assumption that a **HALT.CHECKER** program existed.

The proof may seem a bit mysterious, but it is correct and we must accept the result as true. This is not the same as understanding it, however, and for this, we must also acquire some intuition. One way of viewing the situation is this: The most straightforward way of seeing if a program will halt on input **0** is to run the program and see if it halts. The problem with this approach is that, unless the program has halted, there is no way of knowing if it will halt soon or never halt. The reason for this is that it can take arbitrarily long for a program to halt. For example, if we agree to let n abbreviate any given number, the following program takes at least n steps to compute its value:

```
TO SLOW.n.CONSTANT :X
OP SLOW.n.AUX :X 0
END

TO SLOW.n.AUX :X :K
IF :K = n [OP 1]
OP SLOW.n.AUX :X :K + 1
END.
```

This example is a simple one and we can easily tell how many steps are needed to come to a halt. The **GCD** program,

```
TO GCD :X :Y
IF :Y = 0 [OP :X]
OP GCD :Y REMAINDER :X :Y
END
```

is a bit more difficult to analyze. It is true that **:Y** will give an upper bound on the number of steps before a halt, but the exact number of steps is hard to determine. We can readily imagine programs for which we have no easy determination of the number of steps required to come to a halt. Indeed, our original program **FIB**, for calculating the nth Fibonacci number, probably was surprising in the length of time it took. In general, there is no way of telling how many steps a halting program will take before it halts and there is thus no better way of telling whether or not a program will halt than to run it and wait until it halts. This method obviously fails for nonhalting programs, from which fact we should expect the halting problem to be unsolvable.

A.2 Rice's Theorem

The halting problem is unsolvable. Some properties of programs are solvable. For example, we can effectively decide if the program has 13 lines or not. We can decide if it uses any particular commands or if it localizes all of its variables. Yet many other problems are not solvable. For example, restricting our attention to programs involving natural numbers, we cannot decide if a program halts for all natural numerical inputs even if we know it halts for 0. Just how pervasive is this unsolvability? And, how do our solvable properties differ from the unsolvable ones? Rice's theorem largely answers these questions.

Convention Throughout this section, we restrict our attention to programs possessing a single input variable.

> A property of programs is **nontrivial** if some programs possess the property and some do not.

Example A.2.1 (a) The property of having exactly 13 lines is nontrivial because some programs have exactly 13 lines and some do not.

(b) The property of having exactly one input variable is trivial because we are restricting our attention to those programs that have such. More generally, without this restriction, this would be a nontrivial property.

(c) The halting property of halting on a given input or nonempty set of inputs is nontrivial.

(d) The property of a program's being "correct" after having been tested on a finite set of test data is nontrivial, although this might not be so obvious.

(e) The property of a program's computing a given computable function is nontrivial because, by definition, some program computes it and yet not all programs compute the same function. For a noncomputable function (for example, one that takes programs as inputs and outputs "true" when they halt and "false" otherwise), the property is trivial because no program has it.

> A property of programs is called **extensional** if whenever two programs compute the same function, either both programs have the property or neither does.

Before giving some examples, let us note that all programs can be transformed trivially into functions. Nonfunctions are merely those programs that halt without outputting any value. Such programs can be made functional by replacing every stop command by a call to **CYCLE.FOREVER 0** and inserting such a call at the end of the program. (Through **CATCH** and **THROW**, we can even cycle if the program would otherwise crash through an error.) Hence, we can restrict our attention to programs that compute functions. We shall make this our second convention for this section.

Convention Throughout this section, we assume all programs to calculate functions (not necessarily total, that is, we do not assume they always halt and output a value but only that they output values whenever they halt).

Now we consider some examples of extensional and nonextensional properties:

Example A.2.2 (a) The halting property—that is, the property of halting on some given set of inputs—is extensional.

(b) The property of computing a given computable function is extensional.

(c) The property of computing a function from a given class of computable functions is extensional.

(d) The property of having exactly 13 lines is not extensional because we can always add extra lines that do nothing—for example, we could call the program

**TO DO.NOTHING
END.**

(e) The property of using a specific command is, in general, not extensional.

Nontrivial, extensional properties are undecidable.

Rice's theorem Let P be any nontrivial, extensional property of programs. There is no program that takes programs as inputs and outputs **TRUE** for those programs having P and **FALSE** for the others.

Proof: Suppose, to the contrary, such a program **P.CHECKER** existed. By nontriviality, there are programs **A** and **B** such that **A** has property P and **B** does not. Let a and b abbreviate the texts of these programs. Define

```
TO ANTI.P.CHECKER :X
IF P.CHECKER TEXT "ANTI.P.CHECKER
   [DEFINE "F b]
   [DEFINE "F a]
OUTPUT F :X
END.
```

Now, ask if **ANTI.P.CHECKER** has property P.

If **ANTI.P.CHECKER** has property P, on giving it any input x, it asks if x has property P and, given the affirmative answer, it defines a new function **F**, which computes exactly what **B** does and outputs that result. In other words, if **ANTI.P.CHECKER** has property P, then it computes the same values as **B**. But, P is extensional, and thus **B** has property P as well, contrary to the choice of **B**. Hence, **ANTI.P.CHECKER** does not have property P.

Since **ANTI.P.CHECKER** does not have property P, in running it on any given input x, it first asks and is answered negatively about P. So, it defines a function **F**, which does exactly what **A** does and outputs that value. In other words, since **ANTI.P.CHECKER** does not have property P, it behaves exactly like **A**, which does have property P. Again, the extensionality of P is violated.

We thus arrive at a contradiction, and, since our reasoning was correct, we must reject the assumption that **P.CHECKER** existed.

A. 3 Induction

You should already have some familiarity with induction. You might, however, be used to using induction only in proving simple identities such as

$$\sum_{i=1}^{n} i = \frac{n(n+1)}{2}$$

In the present section, we discuss much more serious applications of induction. We use induction to prove the correctness of several programs. We also prove a couple of other results about programs.

By the limiting results of the previous sections, there is no simple mechanical way of checking if our programs are correct. Thus, if we wish to know if our programs are correct, we must verify them to be so. For recursive procedures, the natural method to use is induction.

Before getting down to business, it should be noted that proofs of program correctness are not of concern only to theoretical computer scientists. Very large programs, too large for us to handle here, often contain many errors that do not show up in testing. For example, one airline's

Very large programs, too large for us to handle here, often contain many errors that do not show up in testing. For example, one airline's reservation program worked beautifully when it was tested in small airports, but when it was put to use at large airports, it started reserving seats for 200 passengers on 90-passenger flights. The alternative to trying the software and seeing if it works is to prove that it does. On large programs, this will have to be done by computer—something that has not been particularly successful yet. Fortunately for us, we have only to prove the correctness of a few simple programs.

Let us begin by recalling the principle of induction.

The **principle of mathematical induction** is the assertion that in order to prove that all natural numbers have a property P, it suffices to prove two things:

1. *Basis*: $P(0)$; that is, 0 has property P.

2. *Induction step*: For any integer k, if $P(k)$ holds, then $P(k + 1)$ also holds.

(For us, the natural numbers begin with 0. Actually, the principle of mathematical induction can begin anywhere and be used to prove that every integer from a given point on has a property in question.)

The standard introductory example of induction is, as we said earlier, the following.

Theorem For any n,

$$1 + 2 + \ldots + n = \frac{n(n+1)}{2}$$

Proof: By induction.

Basis For $n = 0$, the sum is vacuous, so it has value 0, which is $0(0+1)/2$. (If you do not like this, let $n = 1$ be the basis and observe that the sum is 1, which is $1(1+1)/2 = 1 \cdot 2/2$.

Induction Step Assume the result true for $n = k$ and consider the sum of the first $k + 1$ natural numbers:

$$1 + 2 + \cdots + k + (k+1) = k(k+1)/2 + (k+1) \text{ by } P(k)$$
$$= (k+1)(k/2 + 1)$$
$$= (k+1)(k+2)/2$$

and thus $P(k + 1)$.

Thus, we have proven the basis and the induction step, and it follows by the principle of mathematical induction that *P(n)* holds for all n, where *P(n)* is the identity in question.

Exercise A.3.1. Prove by induction:

$$1^2 + 2^2 + \cdots + n^2 = n(n+1)(2n+1)/6$$

We wish to apply induction to show the correctness of programs. When we do this, the predicate $P(n)$ may not be quite so obvious. Determining the right P will be part of the problem. In the simple cases, $P(n)$ will simply assert that the program under consideration on input n will output $f(n)$ for some given mathematical function f. Here is an example.

Theorem The following program correctly computes 2^n:

```
TO EXP.2 :N
IF :N = 0 [OP 1]
OP 2 * EXP.2 :N - 1
END.
```

Proof: The proof is, of course, by induction. Before launching into the proof, we should say something about the ground rules of proving such things. Since the time of the ancient Greeks, it has been recognized that proofs always rest on assumptions; nothing can be proven outright. What are our assumptions? Firstly, we assume mathematical results, in this case the laws of exponents. In particular,

$$2^{n+1} = 2^n \cdot 2^1 = 2^n \cdot 2.$$

Secondly, we assume that the Logo primitives do exactly what they are intended to do. Our job is to prove that our program computes what we say it does, assuming that everything else is done correctly. Moreover, at this level, we are allowed to assume a perfect computer, which will handle all possible n. In later courses, you will come across such mundane limitations as are given by the fact that real machines have finite memories.

Now, let us get down to the actual proof.

Basis **EXP.2 0** immediately outputs 1 and stops. Since 2^0 is 1, it follows that **EXP.2 0** computes the correct value.

Induction Step Assume $P(k)$: **EXP.2** k outputs 2^k. Look at the computation of **EXP.2** $k+1$. First, the machine checks that $k + 1$ is not 0 and decides to call **EXP.2** k, multiply this by 2, and output the result. By $P(k)$, this means the machine will compute $2 \cdot 2^k$ and output this result. But, by the above cited instance of the law of exponents, this means the machine will output 2^{k+1}, which is the correct value.

Exercise A.3.2. Prove the correctness of the following program, which purports to calculate $n!$:

```
TO FACTORIAL :N
IF :N = 0 [OP 1]
OP :N * FACTORIAL :N - 1
END.
```

As mentioned a few lines back, sometimes there is an art to choosing P. The following example of recursion on a list exemplifies this.

Theorem The following program correctly reverses a list:

```
TO REVERSE :LIST
IF :LIST = [ ] [OP :LIST]
OP FPUT LAST :LIST REVERSE BL :LIST
END
```

Proof: This program does not take numerical input. How are we to do the induction? There are two answers. In higher mathematics, we would allow a principle of induction on lists. Less abstractly, we can here just induct on the lengths of the lists. This means we choose as $P(n)$ this assertion: For all lists of length n, **REVERSE** correctly reverses the lists.

Basis For $n = 0$, the result is clear: The only list of length 0 is [ ] and **REVERSE** [ ] outputs [ ], which is correct.

Induction Step Assume $P(k)$, that is, that the program **REVERSE** correctly reverses all lists of length k. Let $[a_1\ a_2\ ...\ a_k\ a_{k+1}]$ denote an arbitrary list of length $k + 1$ and see what **REVERSE** does to it:

REVERSE $[a_1\ a_2\ ...\ a_k\ a_{k+1}]$ simplifies to **FPUT** a_{k+1}
REVERSE $[a_1\ a_2\ ...\ a_k]$, which, by $P(k)$,

simplifies to

FPUT a_{k+1} $[a_k\ ...\ a_1\ a_2]$

which simplifies to

$[a_{k+1}\ a_k\ ...\ a_1\ a_2]$

This is the reverse of the original list.

Exercise A.3.3. The following program correctly determines the length of a list:

```
TO LENGTH :LIST
IF :LIST = [ ] [OP 0]
OP 1 + LENGTH BL :LIST
END.
```

a. State $P(n)$ in such a way that "for all n, $P(n)$" asserts the correctness of the program.
b. Prove the correctness of the program by induction on n.

Sometimes it is not clear what we want to prove. For example, suppose a student turns in the following program:

```
TO MYSTERY.SUM :N
IF :N = 0 [OP 0]
OP 1 / (:N * (:N + 1)) + MYSTERY.SUM :N - 1
END.
```

The teacher wants to see what this program does and prove that it does what the student thinks it does. It is not hard to see that the program calculates the sum

$$\sum_{i=1}^{n} \frac{1}{(i+1)} = 1/(1 \bullet 2) + 1/(2 \bullet 3) + \cdots + 1/(n \bullet (n+1))$$

However, this is not the best possible description of the sum. To get a better one, our teacher can make a little table of values of the program **MYSTERY.SUM**:

n	0	1	2	3	4	5
Sum:	0	1/2	2/3	3/4	4/5	5/6

The obvious conjecture is that **MYSTERY.SUM N** outputs $n/(n+1)$. This is readily verified by induction. You may wish to carry out the details of such before doing the next two exercises.

Exercise A.3.4. Find a simple expression (without using the sigma notation) for the sum $1 \bullet 1! + 2 \bullet 2! + \cdots + n \bullet n!$ and prove that your expression is correct.

Exercise A.3.5. (Difficult). The following program calculates a polynomial:

```
TO GUESS.WHAT :N
IF :N = 0 [OP 0]
OP :N * (:N - 1) + GUESS.WHAT :N - 1
END.
```

What polynomial does it calculate? Prove the correctness of your guess. (*Hint*: If $p(x)$ is a polynomial and $p(a) = 0$, then $p(x) = (x-a)q(x)$ for some polynomial $q(x)$.)

Appendix B
Glossary

arity The number of input values required by a function or procedure.

binary function A function that requires exactly two input values.

binding powers A pair of integers associated with functions and procedures that are used to decide which input values belong to each function or procedure in a Logo expression or statement.

body The body of a user-defined function or procedure is the sequence of statements to be executed when the function or procedure is called.

boundary type A property of the graphics screen. See the **FENCE**, **WRAP**, and **WINDOW** procedures.

buffer A part of memory used to temporarily store text or other data.

combination A function or procedure name and zero or more expressions describing the input values to the function or procedure.

command A request to perform an action that modifies something in the computing environment. It is also called a procedure.

constructor A function that returns a single object constructed from two or more inputs. Examples **FPUT, LPUT, WORD, LIST**

control Rules that determine the order of execution of statements in a program.

cursor A rectangle region of light that indicates a character position on the textscreen.

delimiter Any character that serves as a word separator. In Apple Logo, +, *, -, /, =, <, >, [,], (,), space, and RETURN are delimiters.

dynamic scoping rule The scoping rule used by Logo in determining the value of a variable reference.

editor A program used to assist in creating and modifying programs.

evaluator A program that evaluates expressions according to a collection of simplification rules.

expression A sequence of words and lists that simplify to a word or list.

frame A component in a level diagram containing the body of a user-defined function or procedure and the variable bindings to be used for a particular call to the function or procedure.

function A Logo primitive or user-defined program that returns (outputs) a value without modifying anything else in the computing environment.

functional programming A style of programming based soley on defining functions to return values. That is, the program is designed to return a single value without modifying anything in the computing environment.

global variable A variable that exists at Logo's toplevel.

graphical object An object displayed on the graphics screen that is described by a collection of properties such as size, location, and color.

graphics screen The plane on which the turtle moves.

heading The direction the turtle is facing.

infix notation Expressions where operation names are placed between the input expressions.

input expression An expression describing an input value.

input value A word or list required for the execution of a function or procedure.

input variable A variable denoting a required input value for a function or procedure.

level diagram A graphical model used in describing the execution of programs.

list A collection of words and lists enclosed by square brackets ([and]).

local variable An input variable, or a variable created by a **LOCAL** statement.

object A word or a list.

parser A module of the reader that builds the representation of an expression from the sequence of tokens making up the expression.

penstate A list denoting the turtle's current pen color and pen position.

position property A list denoting the turtle's current location.

postfix notation Expressions where operation names are placed to the right of the inputs.

predicate A function that returns **TRUE** or **FALSE**.

prefix notation Expressions where operation names are placed to the left of the inputs.

primary operation The last function to be executed in a Logo expression.

primitive One of the functions or procedures built into the Logo programming language.

procedure A request to perform an action that modifies something in the computing environment. It is also called a command.

procedural programming A style of programming based on the modification of the programming environment. It usually involves the modification of variables using a **MAKE** statement.

prompt A character or message used to signal the user that input from the keyboard is needed.

property list A collection of name value pairs associated with a word. The pair describes the property name and current value for the property. This facility is built into the Logo language. See the **PPROP, GPROP, REMPROP, PLIST** primitives.

reader A program that reads in text and returns a structure suitable for further processing.

recursion A form of control where a function or procedure calls itself.

scanner The part of a reader that reads a sequence of characters making up a token.

scoping rule A rule describing what value to use when a variable is referenced.

selector A function that returns a piece of its input value. For example, **FIRST [A B C D]** returns **A**.

simplification rules A set of rules that specify how an expression or program is to be executed.

simplifier A program that evaluates expressions according to a collection of simplification rules.

statement The basic structural unit in the Logo language. It consists of the name of a procedure and zero or more expressions describing the input values to be used by the procedure.

symbol table A collection of name value pairs. For example, a telephone directory associates a phone number with the name of a person.

tail recursion A form of recursion where the last statement executed in the body of a function or procedure is either an **output** statement or a recusive call to itself.

text screen The object used to display text.

token The representation of an atomic object used by a parser.

toplevel The module performing Logo's interactive loop of (1) displaying the ? prompt, (2) reading a statement, and (3) performing the desired action.

truth value One of the words **TRUE** or **FALSE**.

variable A word preceded by a :. *See* input variable, global variable, local variable.

variable binding The association of a value to a variable.

visibility property The word **TRUE** if the turtle is visible, the word **FALSE** if it is not visible.

word A sequence of characters.

workspace The part of the computer's memory that stores user-defined functions, procedures, and global variables.

Appendix C
Using Apple Logo II

This appendix describes the differences encountered if Apple Logo II is used with the text.

C.1 Delimiting Characters

The characters, +, -, *, =, <, >, [,], (,) are delimiters in Apple Logo II. The / character is not a delimiter. This means that some expressions that are meaningful when using Apple Logo will not be meaningful in Apple Logo II. For example, the expression **2/3** will not be meaningful. Spaces must be provided on both sides of the / character to denote 2 divided by 3.

C.2 The Intquotient Function

Quotient 3 2 evaluates to 1.5 rather than 1. The **quotient** primitive is the prefix equivalent of division in Apple Logo II. The Apple Logo II function named **intquotient** is equivalent to the Chapter 1 definition of **quotient**. All references to **quotient** in the text should be replaced by **intquotient**.

C.3 Binding Power Table for Apple Logo II

The binding power table (page 17) in Chapter 1 reflects the implementation of Apple Logo. The table that should be used with Apple Logo II is:

Operation	Binding Power	
	left	right
*	7	8
/	7	8
+	5	6
-	5	6
=	3	4
<	3	4
>	3	4
All prefix operations	0	2

For example, the simplification of **FIRST 27 = 3** is

FIRST 27 = 3
=> FIRST "FALSE
=> "F

in Apple Logo II, while the simplification of the same expression is

FIRST 27 = 3
=> 2 = 3
=> "FALSE

in Apple Logo.

C.4 The Editor

The editor is entered just like in Apple Logo, but the editing commands in Apple Logo II differ significantly. The editor is entered by one of the following:

EDIT
EDIT *name*
EDIT *namelist*

In Apple Logo II a file may be created and edited by entering the editor with

EDITFILE *file name*

Exiting the **EDITOR** and returning control to the **TOPLEVEL** module is done by typing one of the following key combinations:

⌘ | A | returns you to Logo after adding all definitions on the edit screen to the **workspace** dictionary. If an operation by the same name already exists, it is replaced by the one contained in the edit screen.

⌘ | esc | returns you to Logo without making any additions or modifications to the **workspace** dictionary.

There are ten editing commands that are used to move the cursor:

⌘ | < | moves the cursor to the beginning of the line.

⌘ | > | moves the cursor to the end of the line.

⌘ | ← | moves the cursor to the left one word.

⌘ | → | moves the cursor to the right one word.

| → | moves the cursor forward (right) one character.

| ← | moves the cursor back (left) one character.

| ↓ | moves the cursor to the next line.

| ↑ | moves the cursor to the previous line.

⌘ | ↓ | moves the cursor to the next page.

⌘ | ↑ | moves the cursor to the previous page.

There are four editing commands that are used to delete one or more characters:

354 *Appendix C*

DELETE deletes the character to the left of the cursor.

CONTROL-F deletes the character under the cursor.

CONTROL-Y moves all characters under and to the right of the cursor into the Kill buffer.

CONTROL-X moves all characters on the line containing the cursor into the Kill buffer.

There is an additional command that provides a paste feature:

CONTROL-R inserts a copy of the text in the Kill buffer at the current cursor position.

C.5 The File System

A portion of the workspace may be saved using the **SAVEL** procedure:

SAVEL *listofprocedurenames filename*

Data files may be created and read by programs.

C.6 The Turtle

The pen position and pen color are regarded as two separate properties in Apple Logo II. There is no **SETPEN** procedure in Apple Logo II. The pen position is modified only by the **PENUP, PENDOWN, PENERASE,** and **PENREVERSE** procedures. This means that the properties of a turtle are described as a five element list:

[*position heading penposition pencolor visibility*]

Appendix D
Using IBM Logo

This appendix describes the differences encountered if IBM Logo is used with the text.

D.1 Delimiting Characters

The characters, =, <, >, +, -, *, /, [,], (, and) are delimiters in IBM Logo. There is no difference between IBM Logo and Apple Logo as far as delimiting symbols are concerned.

D.2 The Quotient Function

Quotient 3 2 evaluates to 1.5 rather than 1. The **quotient** primitive is the prefix equivalent of division in IBM Logo. In IBM Logo the combination **int quotient** *dividend divisor* is equivalent to **quotient** *dividend divisor* when *dividend* and *divisor* are integers. All references to **quotient** in the text should be replaced by **intquotient**.

D.3 Binding Power Table for IBM Logo

The binding power table (page 17) in Chapter 1 reflects the implementation of Apple Logo. The table that should be used with IBM Logo is:

Operation	Binding Power	
	left	right
*	7	8
/	7	8
+	5	6
-	5	6
<	3	4
>	3	4
=	1	2
All prefix operations	0	4

The difference between IBM Logo and Apple Logo is that = is given binding powers lower than < and >. The difference in values of expressions between these two versions is minimal. For example, the simplification of "FALSE = 2 > 3 is

```
"FALSE  = 2 > 3
   => "FALSE = "FALSE
   => "TRUE
```

in IBM Logo, while the simplification of the same expression in Apple Logo will result in an error:

```
"FALSE  = 2 > 3
   => "FALSE > 3
```

D.4 The Editor

The editor is entered just like in Apple Logo, but the editing commands in IBM Logo differ significantly. The editor is entered by one of the following:

EDIT
EDIT *name*
EDIT *namelist*

In IBM Logo a file may be created and edited by entering the editor with

EDITFILE *file name*

Exiting the **EDITOR** and returning control to the **TOPLEVEL** module is done by typing one of the following control characters:

ESC-KEY returns you to Logo after adding all definitions on the edit screen to the **WORKSPACE** dictionary. If an operation by the same name already exists, it is replaced by the one contained in the edit screen.

CTRL-BREAK-KEY returns you to Logo without making any additions or modifications to the **WORKSPACE** dictionary. Outside the editor, this combination interrupts and stops a running procedure.

Here are other editing commands that are used to move the cursor:

| ← | moves the cursor to the beginning of the line.

| → | moves the cursor to the end of the line.

| ← | moves the cursor to the left one character.

| → | moves the cursor to the right one character.

| ↓ | moves the cursor to the next line.

| ↑ | moves the cursor to the previous line.

PGUP moves the cursor to the previous page.

END moves the cursor to the end of the current page.

HOME moves the cursor to the top of the current page.

There are three editing commands that are used to delete one or more characters:

DEL deletes the character under the cursor.

$\boxed{\leftarrow}$ deletes the character to the left of the cursor.

CONTROL $\boxed{\rightarrow}$ moves all characters under and to the right of the cursor into the Kill buffer.

There is an additional command that provides a paste feature:

CONTROL $\boxed{\leftarrow}$ inserts a copy of the text in the Kill buffer at the current cursor position.

D.5 The File System

The **CATALOG** procedure is named **DIR** in IBM Logo. **DIR** takes a file specification as an optional argument. A file specification is of the form

disk drive: filename.extension

The file specification must include the extension when using the **ERASEFILE** procedure.

A portion of the workspace may be saved using the **SAVEL** procedure:

SAVEL *list of procedure names file name*

Data files may be created and read in IBM Logo.

D.6 The Turtle

The penstate property consists of a list of three elements denoting the pen position, pen color, and the pallette. The **SETPEN** procedure requires a list of three elements in IBM Logo. The **PEN** function returns a list of three elements representing the pen state.

Background colors are:

black	0	gray	8
blue	1	light blue	9
green	2	light green	10
cyan	3	light cyan	11
red	4	light red	12
magenta	5	light magenta	13
brown	6	yellow	14
white	7	white	15

Pencolors:

palette	0	palette	1
cyan	1	green	1
magenta	2	red	2
white	3	brown	3

Appendix E
Logo Primitives For Apple Logo, Apple Logo II, IBM Logo

Unless otherwise noted, primitives exist in all three versions of Logo. The following abbreviations are used to denote which versions of Logo support the primitive.

    APP - Apple Logo
    AP2 - Apple Logo II
    IBM - IBM Logo

E.1 Graphics Primitives

BACK(BK) *num*
> moves the turtle back *num* units.

BACKGROUND(BG)
> returns a number that specifies the background color.

CLEAN
> clears the graphics screen without moving the turtle.

CLEARSCREEN(CS)
> clears the graphics screen and puts the turtle at the home position.

DOT *position*
> places a dot on the screen at *position*.

DOTP *position*
> AP2 only: returns **TRUE** if there is a dot at *position*.

Graphics Primitives 361

FENCE
causes Logo to signal an error message whenever the turtle is sent outside the screen boundaries.

FILL
    AP2,
    IBM only: fills closed shape with current pen color.

FORWARD(FD) *num*
moves the turtle forward *num* units.

HEADING
returns the turtle's heading in degrees.

HIDETURTLE(HT)
makes the turtle invisible.

HOME
moves the turtle to [0 0] with heading of 0.

LEFT(LT) *num*
rotates turtle to the left *num* degrees.

PALETTE(PAL)
    IBM only: returns current pallette number (0 or 1).

PEN
returns the current penstate.
    APP: [*penposition color*]
    IBM: [*penposition color palette*]
    AP2:   [*penposition*]

PENCOLOR(PC)
returns a number specifying current pen color.

PENDOWN(PD)
puts the pen in draw mode.

PENERASE(PE)
puts the pen in erase mode.

PENREVERSE(PX)
puts pen in reverse mode.

PENUP (PU)
        puts pen in up mode.

POS
        returns turtle's current position.

RIGHT(RT) *num*
        rotates turtle to the right *num* degrees.

SETBG *num*
        sets the background color.

SETPAL *num*
    IBM only: sets the pallette of colors (0 or 1).

SETPC *num*
        sets the pen color.

SETPEN *penlist*
    APP,
    IBM only: sets pen properties.
    APP: [*penposition color*]
    IBM: [*penposition color palette*]

SETPOS *position*
        moves the turtle to the specified *position*.

SETSHAPE
    IBM only: sets shape of turtle determined by ASCII *num*.

SETX *num*
        sets the *x*-coordinate of the turtle's position to *num*.

SETY *num*
        sets the *y*-coordinate of the turtle's position to *num*.

SHAPE
    IBM only: returns the turtle's current shape.

SHOWNP
        returns **TRUE** if turtle is visible and **FALSE** otherwise.

SHOWTURTLE (ST)
  makes the turtle visible.

SNAP *num*
 IBM only: copies image under turtle onto the turtle's shape and stores it in the ASCII code *num*.

STAMP
 IBM only: stamps a copy of turtle's shape onto the screen.

TOWARDS *position*
  returns the heading from the turtle to *position*.

WINDOW
  puts graphics screen in window mode. The screen is a window on a much larger plane.

WRAP
  puts graphics screen in wrap mode. Each edge of the screen wraps around to the opposite edge. This wrap-around effect is the default behavior.

XCOR
  returns the turtle's x-coordinate as a decimal number.

YCOR
  returns the turtle's y-coordinate as a decimal number.

E.2 Screen Primitives

CAPS
 IBM only: returns **TRUE** if Caps Lock key is disabled and **FALSE** if active.

CLEARTEXT (CT)
  clears the text screen.

364 *Appendix E*

FULLSCREEN (FS)
    devotes the entire CRT to the graphics screen.

MIXEDSCREEN (MS)
    IBM only: allows both text and graphics on the screen.

.SCREEN
    IBM only: returns number of screens being used (1 or 2).

SETCAPS *truth value*
    IBM only: **TRUE** disables Caps Lock key and **FALSE** enables it.

SETCURSOR *position*
    moves the cursor to specified position ([*column row*]).

.SETSCREEN *num*
    IBM only: sets the number of screens (1 or 2).

SETTC *colorlist*
    IBM only: sets text colors to colorlist ([*foreground background*]).

SETTEXT *num*
    IBM only: sets the text portion of graphics screen to the number of lines specified starting from the bottom.

SETWIDTH *num*
    AP2,
    IBM only: sets the width of the screen to *num* characters per line.

SPLITSCREEN
    APP,
    AP2 only: displays both text and graphics screens.

TEXTCOLOR (TC)
    IBM only: returns a list of the numbers for current text colors.

TEXTSCREEN
    devotes the entire CRT to the text screen.

WIDTH
    AP2,
    IBM only: returns current width of screen.

E.3 Math Primitives

ARCTAN *num num*

takes two numeric inputs, *x* and *y*, and returns the degree measure of the angle whose tangent is *x/y*. The quadrant of the angle is determined by the signs of the inputs.

COS *num*

returns the cosine of *num* degrees.

DIFFERENCE *num1 num2*
    AP2,
    IBM only: returns the difference, *num1 - num2*.

EFORM *num1 num2*
    IBM only: returns *num1* in scientific notation, with *num2* digits.

EXP *num*
    IBM only: returns **E** to the power of *num*.

FORM *num int1 int2*
    AP2,
    IBM only: returns *num* with *int1* digits before the decimal point and *int2* digits after the decimal point.

INT *num*

returns the integer part of *num*.

INTQUOTIENT *num num*
    AP2 only: returns the integer quotient of the inputs.

LN *num*
    IBM only: returns the natural log of *num*.

PI
    IBM only: returns an approximation to pi.

POWER *num1 num2*
    IBM only: returns *num1* raised to the *num2* power.

PRECISION
    IBM only: returns the number of significant digits a number is rounded to when used by Logo.

PRODUCT *num ... num*
    returns the product of the inputs (default number of inputs is 2).

QUOTIENT *num num*
    APP: returns the integer quotient of the two numbers. If the inputs are not integers, it first truncates them.
    AP2, IBM: returns the quotient of the two numbers.

RANDOM *num*
    returns a random integer from 0 to *num* - 1.

REMAINDER *num1 num2*
    returns *num1* modulo *num2*. If the inputs are not integers, it first truncates them.

RERANDOM
    re-initializes the seed for the random number generator.

ROUND *num*
    returns the nearest integer to *num*.

SETPRECISION *num*
    IBM only: sets current precision of numbers to *num*.

SIN *num*
    returns the sine of *num* degrees.

SQRT *num*
    returns the positive square root of *num*.

SUM *num ... num*
    returns the sum of its inputs.

E.4 Word and List Primitives

ASCII *char*

        returns the ASCII code of *char*.

BEFOREP *word word*
    AP2 only: returns **TRUE** if first word precedes the second alphabetically.

BUTFIRST(BF) *object*

        if input is a list, a list containing all but the first element is returned. If input is a word, a word containing all but the first character is returned.

BUTLAST(BL) *object*

        if input is a list, a list containing all but the last element is returned. If input is a word, a word containing all but the last character is returned.

CHAR *num*

        returns the character whose ASCII code is *num*.

COUNT *list*

        returns the number of items in the list.

EMPTYP *object*

        returns **TRUE** if the word or list is empty and **FALSE** otherwise.

EQUALP *object object*

        returns **TRUE** if the two items are the same, **FALSE** otherwise.

FIRST *object*

        if input is a list, the first element is returned. If input is a word, the first character is returned.

FPUT *object list*

        returns a list constructed by adding *object* onto the front of *list*.

ITEM *num list*
                      returns the *num*th item in *list*.

LAST *object*
                      if input is a list, the last element is returned. If input is a word, the last character is returned.

LIST *object ... object*
                      returns a list of the input objects.

LISTP *object*
                      returns **TRUE** if *object* is a list, **FALSE** otherwise.

LOWERCASE *word*
    AP2 only: returns the lower case equivalent of *word*.

LPUT *object list*
                      returns a list constructed by adding *object* at the end of *list*.

MEMBER *object list*
MEMBER *word1 word2*
    AP2 only: returns that part of *list* (*word2*) beginning with the first occurrence of *object* (*word1*).

MEMBERP *object list*
                      returns **TRUE** if *object* is a member of *list* and **FALSE** otherwise.

NUMBERP *object*
                      returns **TRUE** if *object* is a number, **FALSE** otherwise.

PARSE *word*
    AP2 only: returns a list obtained from parsing *word*.

SENTENCE(SE) *object ... object*
                      if inputs are all lists, they are appended into a single list. If any inputs are words, they are regarded as one-element lists in performing this operation.

UPPERCASE *word*
    AP2 only: returns the upper case equivalent of *word*.

WORD *word ... word*
> returns a word constructed by concatenating the input words.

WORDP *object*
> returns **TRUE** if its input is a word, **FALSE** otherwise.

E.5 Property Lists

ERPROPS
> AP2 only: erases all properties from the workspace.

GPROP *name prop*
> returns the value of the property *prop* of the word *name* or the empty list.

PLIST *name*
> returns the property list of *name*.

PPROP *name prop val*
> takes a name, a property, and a value as inputs and associates that property value to the name.

PPS
> displays the properties of everthing in the workspace.

PPS *packagename*
PPS *packagenamelist*
> APP,
> IBM only: displays the properties of everything in the specified package(s).

REMPROP *name prop*
> removes the specified property from *name*.

E.6 Workspace Management

BURY *packagename*
    APP,
    IBM only: buries function, procedure, and variable names in specified package. The following procedures act on everything in the workspace except buried objects unless package name is specified: **ERALL, ERNS, ERPS, POALL, PONS, POPS, POTS,** and **SAVE**.

BURY *name*
BURY *namelist*
    AP2 only: buries function and procedure names specified. The following procedures act on everything in the workspace except buried objects unless package name is specified: **ERALL, ERNS, ERPS, POALL, PONS, POPS, POTS,** and **SAVE**.

BURYALL
    AP2 only: buries all function, procedure, and variable names in the workspace.

BURYNAME *name*
BURYNAME *namelist*
    AP2 only: buries specified variable names.

ERALL
ERALL *packagename*
ERALL *packagenamelist*
    APP,
    IBM only: if no input is provided, all (nonburied) functions, procedures, and variables are erased from the workspace. If input is provided, then only those names in specified package(s) are erased.

ERALL
    AP2 only: erases all functions, procedures, variables, and properties from the workspace.

ERASE(ER) *name*
ERASE(ER) *namelist*
> erases named functions and procedures from the workspace.

ERN *name*
ERN *namelist*
> erases named variables from the workspace.

ERNS
> erases all (unburied) names from the workspace.

ERNS *packagename*
ERNS *packagenamelist*
> APP,
> IBM only: erases all (unburied) names in specified package(s) from the workspace.

ERPS
> erases all (unburied) functions and procedures from the workspace.

ERPS *packagename*
ERPS *packagenamelist*
> APP,
> IBM only: erases all (unburied) functions and procedures in specified package(s) from the workspace.

NODES
> returns the number of currently free nodes. This is a measure of how much storage is available in the workspace.

PACKAGE *packagename name*
PACKAGE *packagename namelist*
> APP,
> IBM only: puts named functions and procedures into a package named *packagename*.

PKGALL *packagename*
    APP,
    IBM only: causes all previously unpackaged procedures and variables currently in the workspace to be put in the package named *packagename*.

PO *name*
PO *namelist*
    APP,
    IBM only: prints the definition(s) of the named functions and procedures.

POALL
    prints the definitions of all (unburied) functions, procedures, and variables.

POALL *packagename*
POALL *packagenamelist*
    APP,
    IBM only: prints the definitions of all functions, procedures, and variables in named package(s).

PON *name*
PON *namelist*
    AP2 only: prints the name(s) and value(s) of named variables.

PONS
    prints the names and values of all variables in the workspace.

PONS *packagename*
PONS *packagenamelist*
    APP,
    IBM only: prints the names and values of all variables in the named package(s).

POPS
    prints the definitions of all (unburied) functions and procedures.

POPS *packagename*
POPS *packagenamelist*
    APP,
    IBM only: prints the definitions of all (unburied) functions and procedures in the named package(s).

POT *name*
POT *namelist*
    AP2 only: prints out the title line of named function(s) and procedure(s).

POTS
                prints out the title line(s) of all functions and procedures in the workspace.

POTS *packagename*
POTS *packagenamelist*
    APP,
    IBM only: prints out the title lines of all functions and procedures in the specified package(s).

RECYCLE
                calls the garbage collector.

UNBURY *packagename*
    APP,
    IBM only: unburies all function and procedure names in the specified package.

UNBURY *name*
UNBURY *namelist*
    AP2 only: unburies all function and procedure names specified.

UNBURYALL
    AP2 only: unburies all functions, procedures, and variable names.

UNBURYNAME *name*
UNBURYNAME *namelist*
    AP2 only: unburies specified variable name(s).

E.7 The Outside World

BUTTONP *num*

> returns **TRUE** or **FALSE** depending on whether the button on the indicated paddle (0, 1, 2, 3) is pressed.

KEYP

> returns **TRUE** if a key has been typed but not yet read. Otherwise, **FALSE** is returned.

PADDLE *num*

> returns a number (0-255) depending on the setting of the indicated paddle (0, 1, 2, 3) dial.

PRINT(PR) *object ... object*

> displays the objects on the text screen separated by spaces, and moves cursor to the next line. Lists are displayed without outermost pair of brackets.

READCHAR(RC)

> returns the least recent character in the character buffer, or if empty, waits for an input character.

READCHARS(RCS) *num*
AP2,
IBM only: returns the specified number of characters read by the current file or device (default is keyboard).

READLIST(RL)

> returns a list of words and lists typed in at the keyboard. (It will wait until RETURN is pressed.)

READWORD(RW)
AP2,
IBM only: returns first word in current file or device (default is keyboard).

SHOW *object*

> displays the object on the text screen. Cursor moves to beginning of next line.

TONE *frequency duration*
    IBM only: produces musical note with indicated frequency and duration.

TOOT *frequency duration*
    AP2 only: see TONE.

TYPE *object*
    displays objects on the textscreen with no spaces between them and does not move the cursor to the next line.

E.8 File Management

ALLOPEN
    AP2,
    IBM only: returns a list of devices/files that are currently open.

CATALOG
    APP,
    AP2 only: displays the names of files contained on the currently mounted disk.

CLOSE *filename*
    AP2,
    IBM only: closes a currently opened device/file.

CLOSEALL
    AP2,
    IBM only: closes all currently opened files and devices.

CREATEDIR *directoryname*
    AP2:   creates a directory with given specification.

DIR *file specification*
    IBM only: displays names of files with indicated file specification.

DISK
    APP,
    IBM only: returns a list identifying the disk drive, the slot number of the disk drive, and the volume number of the disk most recently used with CATALOG or SETDISK.

DRIBBLE *filename*
    AP2,
    IBM only: start recording text screen to the device/file.

EDITFILE *filename*
    AP2,
    IBM only: enters Logo editor with the contents of *filename*.

ERASEFILE *filename*
    erases specified file from the disk.

FILELEN *filename*
    AP2,
    IBM only: returns the length of *filename* in bytes.

FILEP *filename*
    AP2,
    IBM only: returns **TRUE** if *filename* exists. Otherwise, **FALSE**.

LOAD *filename*
    loads a file from disk.

LOAD *filename packagename*
    APP,
    IBM only: loads a file from disk into designated package.

LOADPIC *filename*
    AP2,
    IBM only: loads a saved picture onto the graphics screen.

NODRIBBLE
    AP2,
    IBM only: ends dribble process and closes the dribble file.

ONLINE
    AP2 only: returns volume names of all disks on line.

OPEN *filename*
    AP2,
    IBM only: opens a data file.

POFILE *filename*
    AP2,
    IBM only: prints out the contents of the specified file.

PREFIX
    AP2 only: returns the current ProDOS prefix.

PRINTPIC *port*
    AP2 only: prints contents of graphics screen to *port*.

PRINTPIC
    IBM only: prints contents of graphics screen.

READER
    AP2,
    IBM only: outputs the name of current file opened for reading.

READEOFP
    IBM only: outputs **TRUE** if the position of the file being read is the end of the file.

READPOS
    AP2,
    IBM only: returns position in current reader.

RENAME *filename newfilename*
    AP2 only: renames a file.

SAVE *filename*
        saves the contents of the workspace in *filename*.

SAVE *packagename filename*
SAVE *packagelist filename*
    APP only: saves the designated package(s) in *filename*.

SAVEL *name filename*
SAVEL *namelist filename*
    AP2 only: saves specified procedures in file.

SAVEPIC *filename*
    AP2,
    IBM only: saves graphics screen in a file.

.SETCOM *parity databits stopbits*
    IBM only: sets the serial communication line.

SETDISK *drive*
    APP,
    IBM only: sets current disk access to indicated drive, slot number, and volume number. Last two inputs are optional.

SETPREFIX *prefix*
    AP2 only: sets the ProDOS file prefix.

SETREAD *filename*
    AP2,
    IBM only: sets the device/file from which **READCHAR**, **READCHARS**, **READLIST**, and **READWORD** will be read.

SETREADPOS *num*
    AP2,
    IBM only: sets the file position for reading the current file.

SETWRITE *filename*
    AP2,
    IBM only: sets the destination of inputs to **PRINT, TYPE, SHOW, PO,** and **POFILE.**

SETWRITEPOS *num*
    AP2,
    IBM only: sets the file position for writing into the current file.

WRITEOFP
    IBM only: returns **TRUE** if the write pointer is at the end of the file being written.

WRITEPOS
    AP2,
    IBM only: returns position of write pointer in file currently being written to.

WRITER
    AP2,
    IBM only: outputs name of current device/file open for writing.

E.9 Defining and Editing

COPYDEF *name1 name2*
    makes the definition of *name1* the same as that of *name2*.

DEFINE *name deflist*
    defines a function or procedure. The *deflist* is a list whose elements are a list of input variables and a list for each line in the body.

DEFINEDP *name*
    returns **TRUE** if the word is the name of a function or procedure, and **FALSE** otherwise.

EDITFILE *filename*
    AP2,
    IBM only: enters Logo editor with the contents of *filename*.

EDN *name*
EDN *namelist*
    AP2:   enters Logo editor with named variable(s) and corresponding value(s).

EDNS
    APP, AP2,
    IBM: enters Logo editor with all existing variable(s) and corresponding values.

EDNS *package*
EDNS *packagelist*
    enters Logo editor with all existing variable(s) and corresponding values in the specified package(s).

PRIMITIVEP *name*

    returns **TRUE** if *name* is the name of a Logo primitive, and **FALSE** otherwise.

TEXT *name*

    returns a list representing the definition of the named function or procedure.

E.10 Variables

LOCAL *name*

    creates a local variable in the procedure in which it is defined.

MAKE *word object*

    assigns the value of *object* to *word*.

NAME *object word*

    assigns the value of *object* to *word*.

NAMEP *word*

    returns **TRUE** if the word has a value associated with it, and **FALSE** otherwise.

THING *word*

    returns the value associated with *word*.

E.11 Control

CATCH *word list*

    this command, together with its companion **THROW**, provides a means to "abort" procedure executions. **CATCH** takes two inputs: a name and a list of instructions to run. It runs the list, but if at any point while the list is being run, Logo comes across a **THROW** with the same name as the **CATCH**, then

CO
CO *object*

    execution will return immediately to the statement following the **CATCH**.

resumes execution after a **PAUSE**. If an input is present, it becomes the return value of **CO**.

ERROR

returns a six-element list containing information about the most recent error.

GO *word*

transfers control to the line with the label *word*. You can only **GO** to a label within the same procedure.

IF *condition list*

if *condition* is **TRUE**, then the statements in *list* are executed—otherwise, nothing else happens.

IF *condition list1 list2*

if *condition* is **TRUE**, then the statements in *list1* are executed. If the *condition* is **FALSE**, then the statements in *list2* are executed.

IFFALSE (IFF) *list*

executes the statements in *list* if the result of the most recent **TEST** execution is **FALSE**.

IFTRUE (IFT) *list*

executes the statements in *list* if the result of the most recent **TEST** execution is **TRUE**.

LABEL *word*

creates a labeled line for use with **GO**.

OUTPUT (OP) *object*

causes a value to be returned by a function.

PAUSE

interrupts program execution and allows the execution of statements in the current environment.

382   *Appendix E*

                      Execution is resumed by **CO**. **THROW "TOPLEVEL** can be used to abort the pause to return to toplevel.

REPEAT *num list*
                      executes the statement list *num* times.

RUN *list*
                      executes the statements in *list*.

STEP *name*
STEP *namelist*
    AP2 only: allows stepping through named functions and procedures. Logo pauses after each statement and continues when any key is pressed.

STOP
                      causes the current procedure to return control to the calling procedure.

TEST *condition*
                      executes *condition* and remembers the result. Used in conjunction with **IFTRUE** and **IFFALSE**.

THROW *name*
                      transfers control to corresponding **CATCH**.

TRACE *name*
TRACE *namelist*
    AP2 only: displays name and input values of named function or procedure each time it is called. Return information is displayed as well.

UNSTEP *name*
UNSTEP *namelist*
    AP2:   turns off **STEP** mode.

UNTRACE *name*
    AP2 only: turns off **TRACE** mode.

WAIT *num*
                      causes Logo to wait for *num* 60ths-of-a-second before continuing.

E.12 Special Primitives

.AUXDEPOSIT *loc byte*
    AP2 only: stores *byte* at address *loc* in auxillary memory.

.AUXEXAMINE *loc*
    AP2 only: returns the value store at *loc*.

.BLOAD *filename loc*
    AP2,
    IBM only: loads a binary file into memory at *loc*.

.BPT
    APP only: enters the Apple monitor. Logo is resumed by typing **803G** and RETURN.

.BSAVE *filename loc int*
    AP2,
    IBM only: saves *int* consecutive bytes of memory starting at *loc* into a file.

.CALL *loc*
    AP2,
    IBM only: calls a machine language subroutine at *loc*.

.CONTENTS
    returns a list of all symbols that Logo knows about.

.DEPOSIT *loc byte*
    AP2,
    APP only:
    stores *byte* at *loc*.

.DEPOSIT *base offset byte*
    IBM only: stores *byte* at address *base* and *offset*.

.EXAMINE *base offset*
    IBM only: returns the value at memory location specified by *base* and *offset*.

.EXAMINE *loc*

    returns the value stored at memory location *loc*.

.PRINTER *num*
    APP only: if *num* is in the range 1-8, it is the slot to which printer is connected; text will be sent to the printer rather than the screen. If it is in the range 9-15, text goes both to the screen and to slot.

.QUIT
    AP2 only: quits out of Logo, making sure all files are closed.

SCRUNCH
    APP only: returns the current setting of the graphics screen aspect ratio.

.SCRUNCH
    AP2,
    IBM only: returns the current setting of the graphics screen aspect ratio.

SETSCRUNCH *num*
    APP only: changes the vertical scale (the aspect ratio) at which Logo graphics are drawn. The default value for the factor is 0.8.

.SETSCRUNCH *num*
    AP2,
    IBM only: changes the vertical scale (the aspect ratio) at which Logo graphics are drawn. The default value for the factor is 0.8.

Appendix F
Answers to Selected Exercises

(Apple Logo 1.5 is used to perform all computations.)

Chapter 1

Exercise 1.2.3

In Apple Logo, Apple Logo II, and IBM Logo, the \ character (CONTROL Q) is used to indicate that the character immediately following should be treated as an ordinary character.

Example: ?SHOW "2\+3 \-\ 1\ IS\ 4
                2+3 - 1 IS 4

Exercise 1.2.4

Scientific notation is also used to name numbers; eg, **10 000 000 000** will appear as **1.0E10** whereas **0.0000001** will appear as **1.0N7**.

Exercise 1.4.1

a. **QUOTIENT 17 3 + 5**
     => **QUOTIENT 17 8**
     => **2**
(Therefore, quotient is the primary operation.)

d. **(QUOTIENT 17 3) + 5**
     => **5 + 5**
     => **10**
(Therefore, + is the primary operation.)

Exercise 1.5.1

b. 12 - 3 * 2 + 15 / 3
    => 12 - 6 + 15 / 3
    => 6 + 15 / 3
    => 6 + 5.
    => 11.

Exercise 1.5.2

a. 10 - (2 * 4) / (5 - 4) * 2 > 0
    => 10 - 8 / (5 - 4) * 2 > 0
    => 10 - 8 / 1 * 2 > 0
    => 10 - 8. * 2 > 0
    => 10 - 16. > 0
    => -6. > 0
    => "FALSE

c. 1 + ((12 + (15 - 7) * 3) / (4 * (9 - 5) + 2)) = 3
    => 1 + ((12 + 8 * 3) / (4 * (9 - 5) + 2)) = 3
    => 1 + ((12 + 24) / (4 * (9 - 5) + 2)) = 3
    => 1 + (36 / (4 * (9 - 5) + 2)) = 3
    => 1 + (36 / (4 * 4 + 2)) = 3
    => 1 + (36 / (16 + 2)) = 3
    => 1 + (36 / 18) = 3
    => 1 + 2. = 3
    => 3. = 3
    => "TRUE

Exercise 1.5.3

a. REMAINDER SQRT 100 7
    => REMAINDER 10. 7
    => 3

(REMAINDER is the primary operation, SQRT 100, and 7 are the input expressions to remainder.)

c. -REMAINDER QUOTIENT 25 4 5
    => -REMAINDER 6 5
    => -1

(- [negation] is the primary operation, **REMAINDER QUOTIENT 25 4 5** is the input expression to -.)

Exercise 1.5.4

b. **SQRT -QUOTIENT REMAINDER 19 10 4**
   => **SQRT -QUOTIENT 9 4**
   => **SQRT -2**
   ERROR: SQRT DOESN'T LIKE -2 AS INPUT

d. **REMAINDER SQRT 81 (-QUOTIENT 30 11)**
   => **REMAINDER SQRT 81 -2**
   => **REMAINDER 9. -2**
   => **1**

f. **QUOTIENT 40 REMAINDER SQRT 25 - 2**
   => **QUOTIENT 40 REMAINDER 4.79583**
   ERROR: NOT ENOUGH INPUTS TO REMAINDER

Exercise 1.5.5

a. **SQRT 1 + 3 * 5 < 15 / (REMAINDER 23 4) < 14**
   => **SQRT 1 + 15 < 15 / (REMAINDER 23 4) < 14**
   => **SQRT 16 < 15 / (REMAINDER 23 4) < 14**
   => **4. < 15 / (REMAINDER 23 4) < 14**
   => **4. < 15 / 3 < 14**
   => **4. < 5. < 14**
   => **"TRUE < 14**
   ERROR: < DOESN'T LIKE "TRUE AS INPUT

Exercise 1.5.6

a. **2 = 3 = "FALSE**
   => **2 = "FALSE**
   => **"FALSE**

Exercise 1.6.1

a. BUTFIRST BUTFIRST "ZELDA
   => BUTFIRST "ELDA
   => "LDA

d. LAST WORD "BUMBLE "BEE
   => LAST "BUMBLEBEE
   => "E

Exercise 1.7.1

a. FIRST FIRST [[A [B]] C D]
   => FIRST [A [B]]
   => "A

f. SE FIRST [A B C] LIST "A "B
   => SE "A LIST "A "B
   => SE "A [A B]
   => [A A B]

Exercise 1.7.2

a. FPUT 2 + 3 [7 9]
   => FPUT 5 [7 9]
   => [5 7 9]

e. 2 * FIRST [2 3 5] + 4
   ERROR: + DOESN'T LIKE [2 3 5] AS INPUT

Exercise 1.8.1

a. TO FOURTH :WORD
   OUTPUT FIRST BF BF BF :WORD
   END

e. TO DELETE.THIRD :WORD
   OP (WORD FIRST :WORD FIRST BF :WORD BF BF BF :WORD)
   END

Exercise 1.8.2

a. Same as 1.8.1 a).

d. ```
TO REPLACE.ITEM.3 :LIST
OP FPUT FIRST :LIST FPUT FIRST BF :LIST
                         FPUT "THIRD BF BF BF :LIST
END
```

# Chapter 2

## Exercise 2.3.3

```
TO ASSIGNMENT.HEADER :NUMBER :LIST
CLEARTEXT
SETCURSOR [0 0]
TYPE [RONNIE HACKER]
SETCURSOR [0 1]
TYPE [MATH 45, FALL 1986]
SETCURSOR [0 2]
PRINT (SE (WORD "ASSIGNMENT ":) (WORD :NUMBER ",)
         (WORD "PROBLEMS ":) :LIST)
END
```

Exercise 2.5.1

```
TO PO.FIRST :NAME
TYPE "NAME:\ \ \ \ \ \ \
PRINT :NAME
TYPE "INPUT\ VARS:\
PRINT FIRST TEXT :NAME
TYPE "BODY:\ \ \ \ \ \ \
PRINT BF TEXT :NAME
END
```
```

Exercise 2.7.3

```
TO EDITFILE :FILENAME :LIST
SAVE "TEMP
ERPS
LOAD :FILENAME
EDIT :LIST
ERASEFILE :FILENAME
SAVE :FILENAME
ERPS
LOAD "TEMP
ERASEFILE "TEMP
END
```

Chapter 3

Exercise 3.4.1

For exercises a.), b.), and d.) use the following procedure:

```
TO REGULAR.POLYGON :NUM.SIDES :SIDE.MEASURE
HIDETURTLE
FULLSCREEN
REPEAT :NUM.SIDES [FD :SIDE.MEASURE RT 360 / :NUM.SIDES]
END
```

Exercise 3.4.2

```
TO RIGHT1 :NUMBER
SETHEADING (HEADING + :NUMBER)
END
```

Exercise 3.4.4

```
TO PENUP1
SETPEN LIST "PENUP LAST PEN
END

TO SETPC1 :NUMBER
SETPEN LIST FIRST PEN REMAINDER :NUMBER 6
END
```

Exercise 3.4.6

```
TO RANDOM.TURTLE.WALK
REPEAT 100 [ FD RANDOM.INTERVAL 5 15
             RT RANDOM.INTERVAL -45 45]
END

TO RANDOM.INTERVAL :N :M
OP :N + RANDOM :M - :N + 1
END
```

Chapter 4

Exercise 4.5.5

```
TO IS.MEMBERP :OBJ :LIST
IF EMPTYP :LIST [OUTPUT "FALSE]
IF EQUALP :OBJ FIRST :LIST
   [OUTPUT "TRUE]
   [OUTPUT IS.MEMBERP :OBJ BF :LIST]
END
```

Exercise 4.5.6

```
TO MEMBERN :OBJ LIST
IF IS.MEMBERP :OBJ :LIST [OP MEMBERN1 :OBJ :LIST] [OP 0]
END

TO MEMBERN1 :OBJ :LIST
IF EQUALP :OBJ FIRST :LIST
   [OUTPUT 1]
   [OUTPUT 1 + MEMBERN :OBJ BF :LIST]
END
```

Exercise 4.7.4

```
TO SPIRAL.STOP :DISTANCE :INCREMENT ANGLE
HIDETURTLE
IF :DISTANCE > DISTANCE.TO.THE.EDGE [STOP]
FORWARD :DISTANCE
RIGHT :ANGLE
SPIRAL.STOP :DISTANCE + :INCREMENT :INCREMENT :ANGLE
END
```

Exercise 4.8.2

```
TO  CONNECT.AND.POSITION  :POINT.LIST  :TERMINATING.POINT
CONNECT.POINTS  :POINT.LIST
INITIALIZE.THE.TURTLE  :TERMINATING.POINT
END
```

Exercise 4.9.1

```
TO MOVEDOWN :POINTS :NUMBER
IF EMPTYP :POINTS [OUTPUT [ ]]
   [OUTPUT FPUT   LIST FIRST FIRST :POINTS
                       (LAST FIRST :POINTS) - :NUMBER
                  MOVEDOWN BUTFIRST :POINTS :NUMBER
END
```

Exercise 4.9.5

Assumes points are ordered by smallest x-coordinate first.

```
TO SCALE.X :PTS
OP NEW.X :POINTS FIRST FIRST :PTS FIRST LAST :PTS -100 100
END

TO NEW.X :PTS :LOW :HIGH :NEWLOW :NEWHIGH
IF EMPTYP :PTS [OP [ ] ]
   [OP FPUT   (LIST SCALE.PT FIRST FIRST :PTS
                          :LOW
                          :HIGH
                          :NEWLOW
                          :NEWHIGH
                  LAST FIRST :PTS)
          NEW.X BF :PTS :LOW :HIGH :NEWLOW :NEWHIGH]
END

TO SCALE.PT :VAL :LEFT :RIGHT :NEWLEFT :NEWRIGHT
OP :NEW.LEFT + (:VAL - :LEFT) / (:RIGHT - :LEFT)
              * (:NEWRIGHT - :NEWLEFT)
END
```

Exercise 4.9.7

```
TO SCALED.GRAPH :LEFTEND :RIGHTEND :INCREMENT
WINDOW
CONNECT.POINTS SCALE.Y SCALE.X GENERATE.POINTS
               :LEFTEND :RIGHTEND :INCREMENT
END
```

Exercise 4.9.8

```
TO DRAW.GRAPH :POINTS
IF EMPTYP :POINTS [STOP]
CLEARSCREEN
CONNECT.POINTS :POINTS
DRAW.GRAPH MODIFY.GRAPH :POINTS READCHAR
END

TO MODIFY.GRAPH :POINTS :CHR
IF :CHR = "D [OP MOVEDOWN :POINTS 40]
IF :CHR = "U [OP MOVEUP :POINTS 40]
IF :CHR = "L [OP MOVELEFT :POINTS 40]
IF :CHR = "R [OP MOVERIGHT :POINTS 40]
IF :CHR = "Q [OP [ ] ] [OP MODIFY.GRAPH :POINTS :CHR]
END
```

Exercise 4.10.1

```
TO REPLACE.ALL :OBJ :LIST :NEW
IF EMPTYP :LIST [OP :LIST]
IF EQUALP :OBJ FIRST :LIST
   [OP FPUT :NEW REPLACE.ALL :OBJ BF :LIST :NEW]
   [OP FPUT FIRST :LIST REPLACE.ALL :OBJ BF :LIST :NEW]
END
```

Chapter 5

Exercise 5.4.1

a. BUTNTH 3 [A B C D E]
    =>   IF EMPTYP [A B C D E] [OP [ ]]
            IF 3=1 [OP BF [A B C D E]]
                  [OP FPUT FIRST [A B C D E]
                          BUTNTH 3-1 BF [A B C D E]]
    =>   FPUT FIRST [A B C D E] BUTNTH 3-1 BF [A B C D E]
    =>   FPUT "A BUTNTH 2 [B C D E]
    =>   FPUT "A (FPUT FIRST [B C D E] BUTNTH 2-1 BF [B C D E])
    =>   FPUT "A (FPUT "B BUTNTH 1 [C D E]])
    =>   FPUT "A (FPUT "B (BF [C D E]))
    =>   FPUT "A (FPUT "B [D E])
    =>   FPUT "A [B D E]
    =>   [A B D E]

Exercise 5.4.2
a.

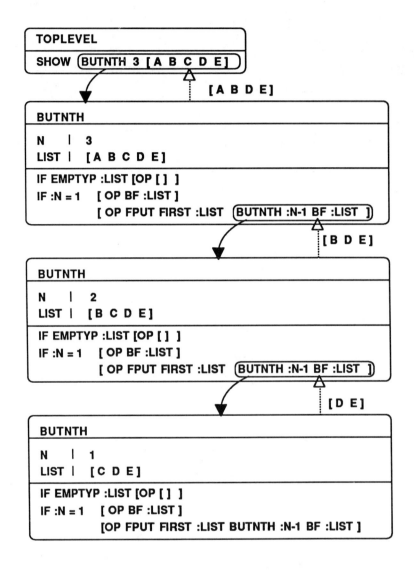

Appendix F   395

Exercise 5.4.6

ALL.BUT [A [A B C] X A B] "A

=>   IF EMPTYP [A [A B C] X A B] [OP [ ]]
     IF EQUALP "A FIRST [A [A B C] X A B]
       [OP ALL.BUT BF [A [A B C] X A B] "A]
       [OP FPUT FIRST [A [A B C] X A B]
              ALL.BUT BF [A [A B C] X A B] "A]
=>   ALL.BUT BF [A [A B C] X A B] "A
=>   ALL.BUT [[A B C] X A B] "A
=>   FPUT FIRST [[A B C] X A B] ALL.BUT BF [[A B C] X A B] "A
=>   FPUT [A B C] ALL.BUT [X A B] "A
=>   FPUT [A B C] (FPUT FIRST [X A B] ALL.BUT BF [X A B] "A)
=>   FPUT [A B C] (FPUT "X ALL.BUT [A B] "A)
=>   FPUT [A B C] (FPUT "X (ALL.BUT BF [A B] "A))
=>   FPUT [A B C] (FPUT "X (ALL.BUT [B] "A))
=>   FPUT [A B C] (FPUT "X (FPUT FIRST [B]
                        ALL.BUT BF [B] "A))
=>   FPUT [A B C] (FPUT "X (FPUT "B ALL.BUT [ ] "A))
=>   FPUT [A B C] (FPUT "X (FPUT "B [ ]))
=>   FPUT [A B C] (FPUT "X [B])
=>   FPUT [A B C] [X B]
=>   [[A B C] X B]

Exercise 5.4.8

```
TO  REPLACE.FIRST :LIST :OBJECT1 :OBJECT2
IF   EMPTYP :LIST [OP [ ] ]
IF   EQUALP FIRST :LIST :OBJECT1
     [OP FPUT :OBJECT2 BF :LIST]
     [OP FPUT FIRST :LIST
           REPLACE.FIRST BF :LIST :OBJECT1 :OBJECT2]
END
```

Exercise 5.5.2

n |0|1|2|3|4|5 |6 |7 |8 |9 |10 |
fib(n) |1|1|2|3|5|8 |13|21|34|55 |89 |
CALLS |1|1|3|5|9|15|25|41|67|109|177|

In general fib(n) requires fib(n-1) + fib(n-2) +1 calls.

Appendix F

Exercise 5.5.3

c.

```
TOPLEVEL
SHOW (ANYWHEREP "A [[ X Y ]  W Z ])
                                        ⋯"FALSE
```

```
ANYWHEREP
W    | A
LIST | [ [ X Y ] W Z ]

IF :LIST = [ ] [ OP "FALSE ]
IF :W = FIRST :LIST [ OP "TRUE ]
IF (WORDP FIRST :LIST) [ OP ANYWHEREP :W BF :LIST ]
IF (ANYWHEREP :W FIRST :LIST) [ OP "TRUE ]
   [ OP (ANYWHEREP :W BF :LIST) ]◁----------
```

"FALSE

```
ANYWHEREP
W    | A
LIST | [ X Y ]

IF :LIST = [ ] [ OP "FALSE ]
IF :W = FIRST :LIST [ OP "TRUE ]
IF (WORDP FIRST :LIST) [ OP (ANYWHEREP :W BF :LIST) ]
IF  ANYWHEREP :W FIRST :LIST [ OP "TRUE ]
    [ OP ANYWHEREP :W BF :LIST ]
```

"FALSE

```
ANYWHEREP
W    | A
LIST | [ Y ]

IF :LIST = [ ] [ OP "FALSE ]
IF :W = FIRST :LIST [ OP "TRUE ]
IF (WORDP FIRST :LIST) [ OP (ANYWHEREP :W BF :LIST) ]
IF  ANYWHEREP :W FIRST :LIST [ OP "TRUE ]
    [ OP ANYWHEREP :W BF :LIST ]
```

"FALSE

```
ANYWHEREP
W    | A
LIST | [ ]

IF :LIST = [ ] [ OP "FALSE ]
IF :W = FIRST :LIST [ OP "TRUE ]
IF (WORDP FIRST :LIST) [ OP ANYWHEREP :W BF :LIST ]
IF  ANYWHEREP :W FIRST :LIST [ OP "TRUE ]
    [ OP ANYWHEREP :W BF :LIST ]
```

"FALSE

ANYWHEREP

| W | A |
| LIST | [W Z] |

IF :LIST = [ ] [OP "FALSE]
IF :W = FIRST :LIST [OP "TRUE]
IF (WORDP FIRST :LIST) [OP ANYWHEREP :W BF :LIST]
IF ANYWHEREP :W FIRST :LIST [OP "TRUE]
   [OP ANYWHEREP :W BF :LIST]

"FALSE

ANYWHEREP

| W | A |
| LIST | [Z] |

IF :LIST = [ ] [OP "FALSE]
IF :W = FIRST :LIST [OP "TRUE]
IF (WORDP FIRST :LIST) [OP ANYWHEREP :W BF :LIST]
IF ANYWHEREP :W FIRST :LIST [OP "TRUE]
   [OP ANYWHEREP :W BF :LIST]

"FALSE

ANYWHEREP

| W | A |
| LIST | [ ] |

IF :LIST = [ ] [OP "FALSE]
IF :W = FIRST :LIST [OP "TRUE]
IF (WORDP FIRST :LIST) [OP ANYWHEREP :W BF :LIST]
IF ANYWHEREP :W FIRST :LIST [OP "TRUE]
   [OP ANYWHEREP :W BF :LIST]

Exercise: 5.5.8

```
TO TREE :TRUNK.LENGTH :ANGLE
IF :TRUNK.LENGTH < 2 [STOP]
FD :TRUNK.LENGTH LEFT :ANGLE
TREE :TRUNK.LENGTH / 2 :ANGLE
RIGHT 2 * :ANGLE
TREE :TRUNK.LENGTH / 2 :ANGLE
LEFT :ANGLE
TREE :TRUNK.LENGTH / 2 :ANGLE
BACK :TRUNK.LENGTH
END
```

Exercise 5.6.5

fib(n - 1) + fib(n - 2) + 1 function calls are required to compute fib(n) using the old fib program. The new version requires only n + 1.

21,891 function calls are required to compute fib 20 using the old fib program. The new version requires only 21 function calls.

Exercise 5.6.6

The number of additions for fib 20 using the old version is 10,945 while the number of additions for fib 20 using the new version is 20.

Exercise 5.7.1

a. Not tail recursive. It isn't recursive.
b. Tail recursive. **IS.MEMBERP** either returns "**FALSE** immediately or the value returned by the recursive call to **IS.MEMBERP**.
d. Not tail recursive. An addition must be performed after returning from the recursive call.
e. Not tail recursive. The **FPUT** operation must be performed after returning from the recursive call.

Chapter 6

Exercise 6.3.1

The numbers **6** and **60** are displayed on the screen. By changing the name of **DISPLAY.SUMS'** input variable to **:LIST**, **ADD** has access to the value of the global variable **:NUMLIST** (the list [10 20 30]).

Exercise 6.3.2

```
TO GENERATE.POINTS :LEFTEND
IF   :LEFTEND > :RIGHTEND [OP [ ] ]
     [OP FPUT   MK.POINT :LEFTEND
                GENERATE.POINTS :LEFTEND + :INCREMENT]
END
```

The values of **:RIGHTEND** and **:INCREMENT** are obtained correctly by the dynamic scoping rule. Since these values never change we do not need to have them as inputs. Since the leftend is changed continually, it must be an input to **GENERATE.POINTS**.

Appendix F

Exercise 6.4.1

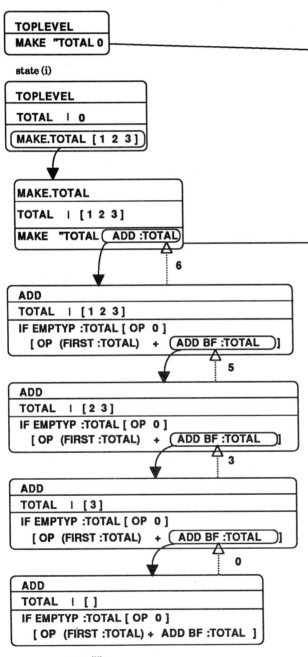

Appendix F **401**

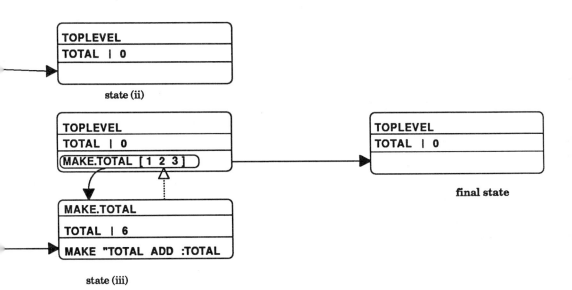

state (ii)

state (iii)

final state

Appendix F

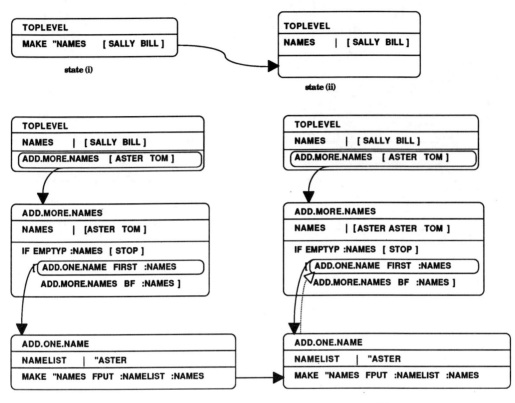

Appendix F **403**

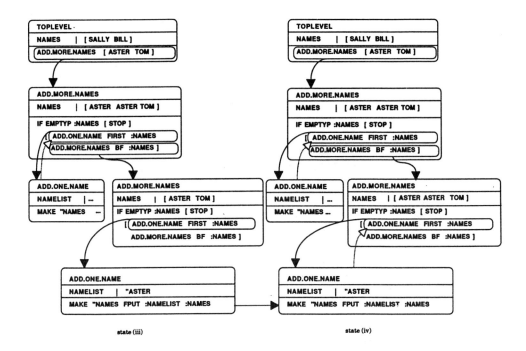

state (iii) state (iv)

The name **ASTER** is repeatedly added to the list associated with **:NAMES** in the most current **ADD.MORE.NAMES** frame.

To fix the problem, rename the input variable:

```
TO  ADD.MORE.NAMES  :NAMELIST
IF  EMPTYP  :NAMELIST  [STOP]
    [ADD.ONE.NAME  FIRST  :NAMELIST
     ADD.MORE.NAMES  BUTFIRST  :NAMELIST ]
END
```

Exercise 6.5.2

```
TO DELETE :NAME
MAKE "DIR DELETE.ENTRY :NAME :DIR
END

TO DELETE.ENTRY :NAME :TBL
IF EMPTYP :TBL [OP [ ] ]
IF :NAME = FIRST FIRST :TBL [OP BF :TBL]
   [OP FPUT FIRST :TBL DELETE.ENTRY :NAME BF :TBL]
END
```

Exercise 6.5.4

```
TO FIND.ALL :FIRST :LAST
OP FIND.ALL1 :FIRST :LAST :DIR
END

TO FIND.ALL1 :FIRST :LAST :TBL
IF EMPTYP :TBL [OP [ ] ]
IF MATCH.INITIALS :FIRST :LAST FIRST FIRST :TBL
   [OP FPUT FIRST :TBL FIND.ALL1 :FIRST :LAST BF :TBL]
   [OP FIND.ALL1 :FIRST :LAST BF :TBL]
END

TO MATCH.INITIALS :FIRST :LAST :NAME
OP MATCH.FIRST :FIRST :NAME MATCH.LAST :LAST :NAME]
END

TO MATCH.FIRST :LETTER :NAME
OP :LETTER = FIRST :NAME
END

TO MATCH.LAST :LAST :NAME
IF "\. = FIRST :NAME [OP :LAST = FIRST BF :NAME]
   [OP MATCH.LAST :LAST BF :NAME]
END
```

Appendix F 405

Exercise 6.6.1

a. TO DISPLAY.INSTRUCTIONS
   PRINT [TO FIND A PHONE NUMBER, PRESS THE 'F' KEY.]
   PRINT [TO DELETE AN ENTRY, PRESS THE 'D' KEY.]
   PRINT [TO UPDATE AN ENTRY, PRESS THE 'U' KEY.]
   PRINT [TO TERMINATE THE PROGRAM,, PRESS THE 'Q' KEY.]
   END

b. TO GET.COMMAND
   TYPE [ENTER COMMAND:\ ]
   OP READCHAR
   END

c. TO READNAME
   TYPE [ENTER NAME:\ ]
   OP FIRST READLIST
   END

Exercise 6.8.1

ANOTHER.AVERAGE uses a local variable to store the sum of the elements while AVERAGE.PROCEDURE creates a global variable to store the sum. ANOTHER.AVERAGE is preferred since the intermediate value will disappear when ANOTHER.AVERAGE returns.

Exercise 6.9.1

TO EQUAL.SETS :S1 :S2
OP AND IS.SUBSET :S1 :S2 IS.SUBSET :S2 :S1
END

Exercise 6.9.3

TO INTERSECTION :S1 :S2
IF IS.EMPTYSET :S1 [OP [ ]]
   [OP MK.INTERSECTION GET.ELEMENT :S1 :S1 :S2]
END

406 *Appendix F*

```
TO MK.INTERSECTION :ELEMENT :SET1 :SET2
IF IS.ELEMENT.OF :ELEMENT :SET2
  [OP ADD.ELEMENT :ELEMENT
       INTERSECTION (DELETE.ELEMENT :ELEMENT :S1) :S2]
  [OP INTERSECTION (DELETE.ELEMENT :ELEMENT :S1) :S2]
END
```

Exercise 6.9.5

```
TO IS.SET :SET
IF NOT LISTP :SET [OP "FALSE] [OP CHK.ELEMENTS :SET]
END

TO CHK.ELEMENTS :SET
IF EMPTYP :SET [OP "TRUE]
IF AND IS.SET.ELEMENT FIRST :SET
       NOT IS.MEMBERP FIRST :SET BF :SET
  [OP CHK.ELEMENTS BF :SET]
  [OP "FALSE]
END

TO IS.SET.ELEMENT :OBJ
IF WORDP :OBJ [OP "TRUE]
IF IS.SET :OBJ [OP "TRUE] [OP "FALSE]
END
```

Chapter 7

Exercise 7.2.2

```
TO SETHEADING :ANGLE
SET.TURTLE MK.TURTLE POS :ANGLE PEN SHOWNP
END

TO SETPC :COLOR
SET.TURTLE MK.TURTLE POS
                     HEADING
                     LIST FIRST PEN :COLOR
                     SHOWNP
END
```

Exercise 7.2.3

```
TO PENCOLOR
OP ITEM 2 ITEM 3 GET.TURTLE
END
```

Exercise 7.3.1

a.
```
TO EXISTS.RESPONDERP
OP NAMEP "CURRENTRESPONDINGTURTLE
END
```

Exercise 7.4.2

a.
```
TO ADD.TURTLE.NAME :NAME
PPROP "TURTLES
       "EXISTINGNAMES
       FPUT :NAME  GPROP "TURTLES "EXISTINGNAMES
END

TO DELETE.TURTLE.NAME :NAME
REMPROP "TURTLES "EXISTINGNAMES
END
```

Exercise 7.4.4

```
TO HATCH :NAME
SET.TURTLE :NAME MK.TURTLE POS HEADING PEN SHOWNP
ADD.TURTLE.NAME :NAME
SENDTO :NAME
END
```

Exercise 7.5.1

```
TO TURTLE.SQUARES :TUR1 :TUR2 :LOC1 :LOC2 :SIZE1 :SIZE2
SENDTO :TUR1 PU SETPOS :LOC1 PD
SENDTO :TUR2 PU SETPOS :LOC2 PD
REPEAT 4 [ SENDTO :TUR1 FD :SIZE1 RT 90
           SENDTO :TUR2 FD :SIZE2 RT 90]
END
```

Exercise 7.5.3

```
TO ALL.SQUARE :TURLIST :SIZE
IF EMPTYP :TURLIST [STOP]
SENDTO FIRST :TURLIST
SQUARE :SIZE
ALL.SQUARE BF :TURLIST :SIZE
END
```

Exercise 7.5.5

```
TO FOUR.BUGS :TUR1 :TUR2 :TUR3 :TUR4
FACE :TUR1 :TUR2
SENDTO :TUR1 FD 1
FOUR.BUGS :TUR2 :TUR3 :TUR4 :TUR1
END

TO FACE :TUR1 :TUR2
LOCAL "TUR2LOC
SENDTO :TUR2
MAKE "TUR2LOC POS
SENDTO :TUR1
SETHEADING TOWARDS :TUR2LOC
END
```

Project 7.1

```
TO CREATE.SQUARE :NAME :SIZE :COLOR
SET.SQUARE :NAME MK.SQUARE [0 0] 0 :SIZE :COLOR
ADD.TO.SQUARENAMES :NAME
END

TO MK.SQUARE :LOC :HDG :SIZE :COLOR
DRAW.SQUARE :LOC :HDG :SIZE :COLOR
OP (LIST :LOC :HDG :SIZE :COLOR)
END
```

(Using property lists)

```
TO SET.SQUARE :NAME :LIST
PPROP :NAME "POSITION ITEM 1 :LIST
PPROP :NAME "HEADING ITEM 2 :LIST
PPROP :NAME "SIZE ITEM 3 :LIST
PPROP :NAME "COLOR ITEM 4 :LIST
END

TO SENDTO.SQUARE :NAME
MAKE "RESPONDER :NAME
END

TO LOC
OP GPROP :RESPONDER "POSITION
END


TO FWD :X
PENUP
SETPOS LOC
SETH HDG
ERASE.SQUARE
FD :X
PPROP :RESPONDER "POSITION POS
DRAW.SQUARE
END
```

Exercise 7.8.1

```
TO HANOI :N
CHECK.HANOI SOLVE.HANOI "A "C "B :N
END
```

Exercise 7.9.1

```
TO DRAWTOWER :NAME
DRAW.SPINDLE LIST GPROP :NAME "LOCATION -75
END

TO DRAW.SPINDLE :LOC
SETPOS :LOC
SETH 0
FD 100
END
```

Exercise 7.9.3

```
TO SOLVE.HANOI.GRAPHICALLY :N
MK.TOWER "A -60 :N
MK.TOWER "B 0 0
MK.TOWER "C 60 0
PERFORM.MOVES SOLVE.HANOI "A "B "C :N
END

TO PERFORM.MOVES :MOVELIST
IF EMPTYP :MOVELIST [STOP]
MOVE.TOP.DISK FIRST :MOVELIST
PERFORM.MOVES BF :MOVELIST
END
```

Chapter 8

Exercise 8.8.2

```
TO RATPRINT :RAT
IF 0 = DENOMINATOR :RAT [PRINT NUMERATOR :RAT]
  [ TYPE NUMERATOR :RAT
    TYPE "/
    PRINT DENOMINATOR :RAT]
END
```

Exercise 8.9.1

a.
```
TO RAT.SUMP :EXP
IF LISTP :EXP [OP "\+ = FIRST :EXP] [OP "FALSE]
END
```

e.
```
TO ARG1 :EXP
OP FIRST BF :EXP
END
```

Exercise 8.9.2

```
TO GCD :A :B
IF :B = 0 [OP :A] [OP GCD :B REMAINDER :A :B]
END
```

Exercise 8.10.2

Use the following function to define **GET.LBP** and **GET.RBP**:

```
TO GET.BP :SYM :N :TABLE
IF EMPTYP :TABLE
   [GEN.RAT.ERROR [NO SUCH OPERATION.] :SYM]
IF :SYM = FIRST FIRST :TABLE
   [OP ITEM :N FIRST :TABLE]
   [OP GET.BP :SYM :N BF :TABLE]
END
```

Exercise 8.12.1

```
TO MK.RATEXP :OPER :ARG1 :ARG2
OP (LIST :OPER :ARG1 :ARG2)
END
```

Chapter 9

Exercise 9.3.3

```
TO BASE.10.TO.C :N10 :BASE
IF :N10 = 0 [OP 0]
   [OP CONVERT.10.TO.C :N10 :BASE]
END

TO CONVERT.10.TO.C :N10 :BASE
IF :N10 = 0 [OP " ]
   [OP WORD CONVERT.10.TO.C QUOTIENT :N10 :BASE :BASE
            BASE.C.REP REMAINDER :N10 :BASE]
END
```

412   *Appendix F*

```
TO BASE.C.REP :DIGIT
OP ITEM :DIGIT + 1 [0 1 2 3 4 5 6 7 8 9 A B C D E F]
END
```

Exercise 9.3.4

```
TO BASE.C.TO.10 :NREP :BASE
IF EMPTYP BF :NREP [OP BASE.10.VALUE :NREP]
  [OP (BASE.10.VALUE LAST :NREP) +
      :BASE * CONVERT.C.TO.10 BL :NREP :BASE]
END

TO BASE.10.VALUE :DIGIT
IF NUMBERP :DIGIT [OP :DIGIT]
IF :DIGIT = "A [OP 10]
IF :DIGIT = "B [OP 11]
IF :DIGIT = "C [OP 12]
IF :DIGIT = "D [OP 13]
IF :DIGIT = "E [OP 14]
IF :DIGIT = "F [OP 15]
END
```

Project 9.2

Step 1

```
TO CALC2
TYPE "CALC2\>
CALC2PRINT CALC2RUN CALC2READ
CALC2
END

TO CALC2PRINT :NUM
PRINT BASE.10.TO.C :NUM 2
END

TO CALC2RUN :LIST
OP RUN :LIST
END
```

```
TO CALC2READ
OP  REPLACE.B2.BY.B10 READLIST
END

TO  REPLACE.B2.BY.B10 :LIST
IF  EMPTYP :LIST [OP [ ] ]
IF  NUMBERP FIRST :LIST
   [OP FPUT  BASE.C.TO.10 FIRST :LIST 2
              REPLACE.B2.BY.B10 BF :LIST]
   [OP FPUT  FIRST :LIST
              REPLACE.B2.BY.B10 BF :LIST]
END
```

Exercise 9.7.2

Start with the following as the toplevel definition:

```
TO NEWTON :F :X0 :TOL
NEWTON1 :F :X0 (NEXTX :F :X0) :TOL
END
```

and define **NEWTON1** and **NEXTX**.

Exercise 9.8.1

```
TO IS.CONSTANT :FUN
OP NUMBERP :FUN
END
```

Using the representation [EXPT *base exponent*] for powers, we define **MK.POWER** by,

```
TO IS.POWER :FUN
IF LISTP :FUN [OP "EXPT = FIRST :LIST]
   [OP "FALSE]
END
```

Exercise 9.8.2

```
TO ARG1 :FUN
OP FIRST BF :FUN
END
```

Index
Logo Primitives Defined in Chapters 1 through 9

Number Primitives

+, -, *, /, =, <, >	12f
NUMBERP	107
QUOTIENT	14
RANDOM	88
REMAINDER	14
SQRT	14

Word and List Primitives

BUTFIRST	28, 31
BUTLAST	28, 31
EMPTYP	107
FIRST	27, 30
FPUT	31
ITEM	223
LAST	27, 31
LIST	32
LISTP	107
LPUT	32
SENTENCE	32
WORD	29
WORDP	106

Property List Primitives

GPROP	234
PLIST	234
PPROP	233
REMPROP	233

Control Primitives

CATCH	269
IF	109
OUTPUT	34, 99
REPEAT	87
RUN	43
STOP	127
THROW	266

Textscreen Primitives

CLEARTEXT	47
CURSOR	48
PRINT	47
SHOW	47
SETCURSOR	48
TYPE	47

Read Primitives

BUTTONP	52
KEYP	52
PADDLE	52
READCHAR	52
READLIST	52

Workspace Primitives

DEFINE	56
DEFINEDP	57

415 *Index*

ERASE	56	SETHEADING	77
ERN	180	SETPOS	78
ERPS	56	SETPC	80
LOCAL	209	SETPEN	80
MAKE	189	SETX	78
NAMEP	202	SETY	78
PO	56	SHOWTURTLE	80
POPS	56	SHOWNP	81
POTS	56	TOWARDS	77
TEXT	56	XCOR	79
THING	202	YCOR	79

CRT Primitives

Graphics Screen Primitives

EDIT	59	BACKGROUND	83
FULLSCREEN	70	CLEAN	83
SPLITSCREEN	70	CLEARSCREEN	83
TEXTSCREEN	70	FENCE	83
		POS	78
		SETBG	83
		WRAP	83
		WINDOW	83

File System Primitives

CATALOG	64
ERASEFILE	64
LOAD	64
SAVE	64

Miscellaneous

AND	105
ASCII	291
ERROR	276
NOT	105
OR	105
TO	34f

Turtle Primitives

Programs Defined in Chapters 1 through 9

BACK	78
FORWARD	78
LEFT	77
HEADING	77
HIDETURTLE	80
HOME	79
PEN	80
PENCOLOR	80
PENDOWN	79
PENERASE	79
PENUP	79
POS	78
RIGHT	77

ABS	110
ADAM	243
ADD	117
ADD.DISK	257
ADD.ELEMENT	215
ADD.ENTRY	202

Index

ADD.GLOBAL	189	DRAWDISK	251
ADD.GLOBAL1	192	DRAWREC	251
ADD.MORE	189	DRAW.SQUARE	86
ADD.NAMES	194	DRAW.GRAPH	136
ADD.NUMS	183	DRAWSQUARES.ONE	174
ADD.PROCEDURE	211	DRAWSQUARES.TWO	174
ADD.TURTLE.NAME	231	EMPTYTOWERP	254
ALL.BUT	157	ERASEDISK	253
ALL.BUT.NTH	157	EVALFUN	140
ANTI.HALT.CHECKER	326	EVE	243
ANTI.P.CHECKER	339	EXP.2	342
ANYWHEREP	162	EXPLORE.GRAPH	136
AVE1	168	FIB	159, 169
AVERAGE	117, 168	FIB1	170
AVERAGE.PROCEDURE	212	FIND	203
BAD.MOVE	248	FIRST.EQUALS.LASTP	108
BANISH	230	FIRST.TWO	54, 60
BASE.10.TO.2	312	FLASH	95
BASE.10.TO.C	313	FRACTIONP	296
BEEP	95	GENERATE.POINTS	135, 140
BETP	106	GENERROR	279
BUTNTH	155	GEN.RAT.ERROR	281
CALC.ANY.TOPLEVEL	315	GEP	106
CALC2	310	GET.CHAR	294
CHECK.HANOI	247	GET.DENOMINATOR	296
CHECK.MOVES	248	GET.HEADING	235
CLEANUP	262	GET.NUMBER	295
CONNECT.ALL	130	GET.NUMERATOR	295
CONNECT.POINTS	130	GET.PENSTATE	235
CONNECT.REST	131	GET.POS	234
CONVERT.10.TO.2	312	GET.RAT	295
CREATE.SQUARE	241	GET.TOPDISK	249
CYCLE.FOREVER	335	GET.VISIBILITY	235
DEFINEPOLY	102	GRAPH	132, 139
DELETE.ELEMENT	215	GROW	96
DELETE.TURTLE.NAME	232	GUESS.WHAT	345
DENOMINATOR	287	HANDLE.ERRORS	280
DESIGN	96	HANOI	247
DIFF	331	HATCH	227, 243
DISKDRAWINIT	251	IS.EMPTYSET	214
DISK.LOC	252	IS.POWER	332
DISK.SIZE	252	IS.SUBSET	218
DISPLAY.SUMS	186	INITIALIZE.TURTLE	130
DOSQUARE	94	LARGE.PROGRAM	268

LARGE.PROGRAM1	270	RATSCAN	293
LOGO	276	REACHED.GOAL	324
LOGOCALC	278	READLINE	290
LOOKUP	203	READLINE.AUX	290
MK.DISK	251	RECTANGLE	89
MK.DISK.NAMES	253	REMOVE.CHAR	295
MK.DISK.STACK	253	REPLACE.ALL	157
MK.EMPTYSET	215	REPLACE.FIRST	157
MK.MOVE	247	REPLACE.FIRST.STATE	325
MK.NEW.STATE.LIST	325	REPLACE.NTH	156
MK.NEW.STATES	325	RETURNP	291
MK.POINT	134, 140	REVERSE	120
MK.RAT	287	SCAN	295, 302
MK.SET	215	SECOND	54, 60
MK.SQUARE	241	SENDTO	228
MK.SUM	332, 324	SET.PRINT	320
MK.TOWER	248, 253	SET.READ	320
MK.TURTLE	223	SET.RUN	320
MODIFY.GRAPH	137	SET.TOPLEVEL	319
MOVECURSOR	50	SET.TURTLE	234
MOVEDISK	252	SHOW.SQ	98
MOVE.TOP.DISK	256	SHOWTHIRD	37
MYSTERY	95, 344	SIGN	122
NTH	115	SIMULATE.NEW.RESPONDER	229
NUMERATOR	287	SOLVE	318
PADDLEPRINT	53	SOLVE.HANOI	246
PEEK.CHAR	294	SOLVE.MCP	324
PEEK.2ND.CHAR	297	SOLVE.MC	327
PERFORM.MOVE	249	SQ	98
POLY	95	SQ1	100
POLY5	103	SQRT1	317
POP.DISK	257	SQUARE	86
POWER	113	SPI	96
RANDOMSPIRAL	127	SPIRAL	125
RAT	280	SPIRAL.IN	126
RATCALC	280	STEP	111
RATCONSTANTP	287	STRIP	111, 124
RATPARSE	298, 302	SUM.SQ	98
RATPARSEPAREN	303	SUM.SQ1	100
RATPLUS	284	TELEPHONE	209
RATREAD	290	TELEPHONE1	206
RATREADRIGHT	298	TELEPHONE2	208
RATRUN	283	TELEPHONE3	208

THIRD	35	counting recursion	112
TOP.DISK	255	CRT primitives	70
TOPLEVEL	42, 275	cursor	2
TOWER.DISKS	254	cursor position	46, 61
TOWER.LOC	254	defining	
TOWER.TOP	255	functions, procedures	34ff
TREE	164	delimiter	4
TROUBLE	63	deviation	307
TURTLES	231	dispersion	306
TWO.SQUARE	236	dynamic scoping rule	184
TWO.TURTLE.SQUARE	236	editing	59
TYPE.DIAGONALLY	49	editing commands	53, 61, 62
UNION	219	edit screen	59, 61
UPDATE.CURRENT.		Editor Module	59
RESPONDER	228	efficiency	166
UNSTABLE	325	END	57
WHERE	98	error handling	276
		escape	260
		to toplevel	264
		catching of	269

General

-, ambiguity of	4	expression	1, 2, 39
arithmetic mean	306	constant	6
arithmetic operations	12ff	combination	6, 7, 39
arity	12	evaluation of	18ff, 40
ascii code	291	extensional property	338
assignment statement	178	Fibonacci sequence	158
base two calculator	310	file	63, 65
binary function	11	file primitives	64
binding, variable	144	File System Module	63
binding powers	16ff	Four Bugs Problem	238
BNF notation	282	frame	144
boundary type	82	frequency table	308
calling chain	142	function	37
catching control	268f	functional programming	210
central tendency	306	global variables	178
class frequency distribution	308	problems with	194
combination (see expression)		graceful return	261
command	1	graphical objects	221
Common Lisp	244	multiple turtles	224
connecting points	128	polygons	242
constant	3, 39	squares	239
simplification of	18, 40	Towers of Hanoi	250
constructor	27	Graphics Screen Module	71, 82
		graphics screen	72, 82, 85

419 Index

graphing	132
greatest common divisor	288
halting problem	335
heading property	73, 75
induction	340
infix notation	11
input expression	7
input stream	51, 52
input value	7
input variable	36
kill buffer	51f, 61
list	3
element of	6
primitives	30ff
level diagrams	142
local variable	178, 208
logical operations	105
Logo environment	41
Logo calculator	277
Logo's toplevel	275
mathematical induction	341
median	306
memory layout	68f
missionaries and cannibals	322
mode	306
models	
level diagrams	142ff
simplification	39
state diagrams	181
multiprocessing	237
nested **IF**s	122
Newton's method	328
nontrivial property	337
number	4
number system calculator	309f
operation table	289, 303
parentheses	8
parser	297
pencolor	75
pen position	75
penstate property	73, 75
polygons as objects	242
position property	73, 75
postfix notation	11
predicate	12, 106
prefix notation	11
prefix operations	13ff
primary operation	7
procedural programming	210
procedure	1, 37
prompt	2
properties of	
disks	251
polygons	242
squares	240
towers	253
turtles	74, 222
property lists	232
primitives	233
range	307
rational-number expressions	
calculator	280
constructor	284
expression	282
parenthesized	301
reader	289
recognizers	285
representation of	287
selectors	284
simplifier	281
Reader Module	49
reading	
from keyboard	52
from game paddles	52
recognizer	104f
recursion	
counting	112f
forever	125
level diagrams	148
list	116f
mutual	218
STOPping	125
tail	173
tree	158
representation independent	285
return values	34

Rice's Theorem	339	toplevel frame	144
scanner	289, 293	Toplevel Module	42
Scheme	244	Towers of Hanoi	244
selector	27	checking solution to	247
set expression		disk functions	
calculator	319	and procedures	251f
representation of	320	graphical solution	250
sets	213	moving disks	256
elements of	213	solution to	246
equality of	216	tower functions	
expressions	320	and procedures	253f
intersection of	216	truth value	4
primitives	213	turtle	
subset of	216	definition of	222
union of	216	multiple	224
side effect	97	as a procedure	242
simplification of prefix		properties	73, 76, 221
expressions	18	programming multiple	
IF	114	turtles	236
infix expressions	19	walk	88
OUTPUT		Turtle Module	73, 76
parenthesized expressions	20, 40	unary operation	11
REPEAT	91f	user interface	206
user defined functions	89, 91	value	8
simplified statement	42	variable	36
Simplifier Module	67	input	36
solving equations	315	global	178
square roots	317	local	178, 208
square world	239	variance	307
standard deviation	307	visibility property	73, 75
state diagram	181	word	3
state tree	322	primitives	27ff
statement	1	Workspace Module	54
statistics application	306	workspace dictionary	54, 55
successive halving	317	workspace primitives	56
symbol table	201		
symbolic differentiation	330		
Text Screen Module	44		
text screen	44f		
primitives	47		
telephone directory application	200		
TO	57		
token	293		